D0143587

Science by the People

Nature, Society, and Culture

Scott Frickel, Series Editor

A sophisticated and wide-ranging sociological literature analyzing nature-society-culture interactions has blossomed in recent decades. This book series provides a platform for showcasing the best of that scholarship: carefully crafted empirical studies of socioenvironmental change and the effects such change has on ecosystems, social institutions, historical processes, and cultural practices.

The series aims for topical and theoretical breadth. Anchored in sociological analyses of the environment, Nature, Society, and Culture is home to studies employing a range of disciplinary and interdisciplinary perspectives and investigating the pressing socioenvironmental questions of our time—from environmental inequality and risk, to the science and politics of climate change and serial disaster, to the environmental causes and consequences of urbanization and war making, and beyond.

Available titles in the Nature, Society, and Culture series:

Science by the People

*Participation, Power, and the Politics of
Environmental Knowledge*

AYA H. KIMURA AND ABBY KINCHY

Rutgers University Press
New Brunswick, Camden, and Newark, New Jersey, and London

Library of Congress Cataloging in Publication Number: 2019005304

A British Cataloging-in-Publication record for this book
is available from the British Library.

Copyright © 2019 by Aya H. Kimura and Abby Kinchy
All rights reserved
No part of this book may be reproduced or utilized in any form or by any means,
electronic or mechanical, or by any information storage and retrieval system, without
written permission from the publisher. Please contact Rutgers University Press, 106
Somerset Street, New Brunswick, NJ 08901. The only exception to this prohibition is
"fair use" as defined by U.S. copyright law.

∞ The paper used in this publication meets the requirements of the American
National Standard for Information Sciences—Permanence of Paper for Printed
Library Materials, ANSI Z39.48-1992.

www.rutgersuniversitypress.org

Manufactured in the United States of America

Contents

Preface

Every book has a genealogy. This one arises in a particular political moment but builds on many years of observations, reflections, and experiences, both professional and personal. We wrote this book when antiscience sentiment seemed to grab the center of power for environmental governance and regulation in the United States, the country where we both work. It was exciting to see people organize to defend public commitments to science and environmental protection, yet we had an uneasy reaction to slogans such as "There are no sides to science" and "Science is facts." In an effort to counter the Trump administration's dismissal of science, the proscience discourse seemed to experience a full swing back to an ideal of science as separate from and uncontaminated by politics and beliefs. But as sociologists who study how people use scientific inquiry to fight for sustainability and environmental justice, we know that science and values are inseparable.

Indeed, the research we did for this book is inevitably influenced by our values as well as our professional and personal positions in society. We talk about our values in chapter 2 of this book, but to increase our transparency about this project, we want to share some of the experiences that shaped the perspectives that we offer in this book. We contributed equally to this book, but here and on the title page, we go in alphabetical order.

Aya H. Kimura:

I did not grow up counting birds or observing butterflies, but I wish I had! As a child in Japan, I thought "hard" sciences were for smart kids and had nothing to do with me. But now that I have two kids, I get excited when I hear about birds and butterflies and how children can participate in data collection with scientists. As a mother who is seriously worried about the future livability of the planet for my children and all children, I know that knowledge about nature and the living creatures with which we cohabit this earth is incredibly important.

As I became concerned with environmental issues, I understood science as an integral part of addressing the problems. My passion for understanding environmental destruction brought me to the School of Forestry and Environmental Studies at Yale for my master's degree. There, I was lucky to become a research assistant to Arun Agrawal, a renowned scholar of commons and indigenous knowledge, and we compiled cases of community management of natural resources. I was also mentored by Michael Dove and Jim Scott, whose works on local knowledge also opened my eyes to different practices besides what is typically legitimated as "science."

I went on to pursue a PhD in environmental sociology, continuing my interests around sustainability and knowledge. For my dissertation, I decided to look at the politics of food and agriculture. More specifically, my topics were fortification and biofortification in developing countries (I did my fieldwork in Indonesia). I examined how Western scientific knowledge was reconfiguring the meaning of food and food insecurity. Following the feminist methodology of "studying up," my focus was on powerful actors such as multinational corporations, Western scientific expertise, international philanthropic organizations, and foreign aid donors that seemed to push a highly circumscribed idea of "good food." This was the central issue of my first book, *Hidden Hunger: Gender and the Politics of Smarter Foods.* But I was also interested in how ordinary people were reclaiming food—in the form of, for instance, food sovereignty movements—so I started to look at the role of science-doing by regular people in my subsequent work. For example, one of my projects examined how regular women engaged with food-quality assurance in collaboration with food producers in Japanese cooperatives.

After the 2011 Fukushima nuclear accident, I began researching citizen scientists testing for radiation (I wrote about the genealogy of this project in the preface for the book *Radiation Brain Moms and Citizen Scientists*, if you are interested in hearing about this). I interviewed many citizen science groups, and I met so many inspiring people. They were measuring a lot of different things and gave information to people who were desperate to know how to protect their families and themselves. Citizen scientists were not doing this for money or even prestige (there was a lot of policing around radiation concern); they were volunteering their time and resources while often raising children, caring for the elderly, and working their "real" jobs.

As I met and read about many more of them, I noticed large variation among citizen scientists, even while they were doing the same thing—measuring radiation. Some were taking the position that data are and should be apolitical, while others were committed to activism. I also started to notice that the pronuclear elites embraced citizen science as an instance of independent and self-responsible citizenry. It hit me that radiation monitoring and management by individual citizens could help circulate the discourse that said, "You can live in a contaminated place as long as *you* are careful."

Citizen science books that I had read did not capture such complexities, but I felt the need to unpack further what was going on. Was this kind of complexity applicable only to radiation? Was this only in Japan? Around the same time at one of our academic conferences, I talked with my graduate school friend Abby, who happened to be mulling over a case of citizen science, this one relating to fracking in the United States. We seemed to have been thinking about similar things.

Abby Kinchy:

Mill Creek is a shallow, narrow stream that winds through forest and farmland before emptying into Stephen Foster Lake, a recreational body of water located in a small state park in Northeastern Pennsylvania. In 1995, I spent the summer walking that creek with another college student identifying benthic macroinvertebrates—bugs that live in streams—in order to help the Bradford County Conservation District assess the quality of the water and to identify possible sources of nonpoint pollution, like runoff from dairy farms. Memories of that internship—the pleasure of getting wet and muddy on hot summer afternoons, exploring woodlands and pastures from the perspective of a creek rather than a road or trail—return to me as I consider the ways that natural gas development is transforming that landscape.

Bradford County, never before an energy extraction region, has experienced the rapid development of shale gas resources since about 2007. Many scientists and environmentalists contend that these activities will degrade water quality both acutely when spills occur and over the long term as the landscape of the watershed is changed. Concern for these changes, combined with my experience monitoring Mill Creek, led me in 2009 to respond to an email announcing an organizing meeting for volunteers interested in learning to monitor streams for the impacts of gas drilling. Though I lived a three-hour drive away, both my mom and I signed up. This was the spark for my research into volunteer water monitoring (which we write about in chapter 3).

I first started thinking about citizen science around 2003 or 2004, as I was beginning my dissertation project and learning about how Mexican corn farmers were testing their crops for genetically engineered material (something we write about in chapter 5). My research focused on how social movements were having an influence on the highly technical realms of biotechnology governance in North America. At the time, I thought of it as "popular biology" or activist-led research, since this was before there was a lot of popular discourse about "citizen science." I later learned more about *community science*, the term used by the organizers of the water monitoring project that I joined.

I really enjoyed participating in that water monitoring group, but over time I started to doubt that our efforts were going to make a difference. Through family friends and fellow researchers, I knew of people who were going through terrible experiences related to gas drilling—illnesses, sick animals, polluted well water, disruptions to their use of farmland, shady interactions with land men, and more. It felt as though we were taking measurements that had little to say about these experiences.

At the same time, I had a couple of experiences where I was invited to collaborate on "citizen science" initiatives that were starting up on my campus. Natural scientists were interested in using volunteers to do innovative research, and computer scientists wanted to develop new ways to aggregate and share citizen-collected data. I decided not to pursue these collaborations because they didn't resonate with my values. I felt that these projects were starting with professional researchers' interests, not concerns for significant public issues or communities affected by environmental problems. Nevertheless, I felt guilty about declining to participate, wondering whether I could have shifted the projects in different directions and whether I was too quick to dismiss their potential to address the causes of environmental problems.

I continued my work as a sociologist, studying the geographic distribution of water monitoring projects and trying to understand the social contexts that shaped their approaches to scientific study. Yet I felt unsure about how to communicate what I was learning in the context of growing public enthusiasm for and media attention to citizen science (this was around the time that the Obama administration was enthusiastically promoting citizen science and held the White House Maker Faire). I shared these thoughts with Aya, and we began to cook up the idea for this book. Perhaps, we agreed, our uneasy feelings about citizen science in each of our projects were pointing us toward a broader set of ideas about the complexities of participatory research in the world today.

* * *

In this book, we are not asking whether citizen science results in good scientific data or how to manage recruitment and retention of volunteers. We both have a background in political and environmental sociology and science and technology studies (STS).

We pursued our PhDs together in the Departments of Sociology and Rural Sociology at the University of Wisconsin–Madison. These disciplines provide important analytic frames to probe citizen science, such as issues of participation, structural inequity, and citizenship. In our fieldwork, we have seen how citizen science can assist groups at the lower end of social hierarchy in gaining voice, legitimating their knowledge and claims, and participating in decision-making processes. How do they do this? What are the tensions and choices involved in doing citizen science that hinder others from achieving this potential? These are the driving questions behind this book.

* * *

We would like to thank our editor, Peter Mickulas at the Rutgers University Press, and series editor, Scott Frickel, for encouraging us to write about these ideas and for their guidance throughout the process of writing and revising this book. Many people have read drafts of various parts of the book. Thanks are due to Gabrielle Hecht, Barbara L. Allen, Gwen Ottinger, Sarah Wylie, Jason Delborne, Jennifer Dodge, Beth Popp Berman, Kendra Smith-Howard, David Hess, and the anonymous reviewers. Their feedback was encouraging and immensely helpful.

The people we interviewed are too numerous to mention individually, but without their generosity of time and knowledge, this book would not have been possible.

Aya H. Kimura wants to thank her colleagues in the sociology department at the University of Hawai'i–Mānoa as well as the Dōshisha University's Global and Area Studies, which has been her home for the field research in Japan. Special thanks are due to Taro Futamura, Toyohisa Ishikawa, Sumito Hatta (who kindly allowed us to use his illustrations, which became the basis for the cover of the book), and Keisuke Amagasa. The research projects in the book were supported financially by grants from the National Science Foundation (award no. 1329111), sociology department; the College of Social Sciences; and the Center for Japanese Studies at the University of Hawai'i–Mānoa.

Abby Kinchy wishes to thank her mom, Sue Kinchy, for accompanying her on the journey into watershed monitoring. She also thanks her colleagues in the STS department at Rensselaer Polytechnic Institute for providing a collegial and encouraging community in which to develop this book. Kirk Jalbert provided invaluable research assistance and ongoing insight about water monitoring and other citizen science in the Marcellus Shale region. Special acknowledgment is due to the graduate students who offered their playful and erudite suggestions for alternatives to the phrase *citizen science*. While in the end we decided not to go with *felagnotology*, *sullegnosis*, *demognocy*, *plebscience*, or *covestigation*, thinking together about these words was one of the joys of working on this book. Thank you, Mara Dicenta, Bucky Stanton, Laura Rabinow, Hined Rafeh, and Mitch Cieminski.

Kinchy's research, discussed in this book, took place over many years, beginning with a dissertation project at the University of Wisconsin–Madison. Over the years, funding for the research has been provided by the National Science Foundation (award nos. 1743138, 1126235, and 0525799), the Social Science Research Council, the New York State Energy Research and Development Authority, and the School of Humanities, Arts, and Social Sciences at Rensselaer.

We both want to express our gratitude to our spouses and children, who have patiently endured our Skype meetings at strange hours and gave us the space we needed to think and write. Nathan Meltz assisted us with the images that appear in this book and created the collage for the cover. Ehito Kimura read chapters and helped improve our prose. Thank you, Nathan, Ehito, Aldo, Emma, and Isato.

Science by the People

1

Environmental Citizen Science

Virtues and Dilemmas

For decades, Chile has been the world's top copper producer, making it the wealthiest country in Latin America. But with mineral wealth comes a tremendous amount of hazardous waste, including tailings, the ground-up materials that remain after copper is removed. Copper tailings can contain lead, arsenic, mercury, and other materials known to threaten human health. Until recently, Chile's government paid little attention to the hundreds of millions of tons of tailings that piled up each year in the mining regions. In 2012, the Ministry of the Environment published procedures for identifying and remediating soils contaminated by mining waste, yet the agency has been slow to carry out the work. Furthermore, there has been little communication with the communities affected by mining waste, who may be exposed to harmful materials in the course of their daily activities.[1]

Sebastian Ureta, a sociologist in Santiago, Chile, has observed his government's efforts to address soil contamination with growing concern. Now skeptical that the Ministry of the Environment has the funds or the will to protect people from copper tailings, Ureta is considering the possibility of an alternative approach—citizen science. He hopes that a low-cost tool or method can be developed for testing soil for contaminants like lead. In the hands of local residents, such tools could inform people of the dangers they face and enable them to reduce their exposures. Perhaps even soil remediation could be accomplished through grassroots research and do-it-yourself (DIY) technologies. More broadly, Ureta hopes that citizen science will empower ordinary people to take leadership in assessing the environmental challenges they face and the possible solutions.

Like Professor Ureta, many environmental professionals, activists, and scholars around the world today consider citizen science to be part of their tool kit for addressing environmental challenges. Getting ordinary people involved in making observations is widely held to be a good way to not only increase public understanding of the environment but also gather large amounts of data that can inform research and problem solving. There are numerous examples. In parts of the United States where "fracking" for oil and natural gas is changing the landscape, volunteers are monitoring acidity, dissolved solids, erosion, biodiversity, and other characteristics of their watersheds. In Japan, dozens of community organizations are analyzing foods for radioactivity resulting from the 2011 nuclear accident in Fukushima. Beyond such grassroots efforts, professional scientists have integrated crowd sourcing and participatory practices into many scientific projects. For instance, the Cornell Lab of Ornithology uses bird observations made by hundreds of thousands of individuals each year.[2]

Citizen science has also become an important public policy issue. Many governments—including the United States, European Union, and Singapore—have expressed desires to encourage and capitalize on citizen science.[3] Some science policy experts have applauded citizen science for making science more participatory, showing the possibility of a scientifically literate society and more equitable engagement between experts and the lay public. Furthermore, many people today are asking whether their organization—a university, nonprofit, activist group, or government agency—should allocate time and resources to citizen science and, if so, how to do it in a way that is fair and beneficial to the volunteers. Weighing on these decisions are some serious critiques. There are those who discount the contributions of nonprofessional researchers, contending that volunteers can't do "real science." Others argue that citizen science exploits the free labor provided by volunteers, primarily to the benefit of big business and cost-cutting government agencies.[4]

This book shares neither the naive enthusiasm for citizen science nor the cynicism that permeates many critiques. As environmental sociologists, we aim to understand how citizen science affects people's lives and environments. Our goal here is to provide a language for talking about the complexity and contradictions within environmental citizen science so we can move beyond generalizations to grapple with the specific dilemmas that participatory research projects commonly face.

Our interest in contradictions and dilemmas arises out of our research projects on the fracking and Fukushima controversies. In these examples, we saw that citizen science was a valuable tool for those concerned with protecting the health of their community and environment. However, we learned that participation in scientific knowledge making did not necessarily enable people to challenge powerful institutions (such as the nuclear industry) or achieve desired outcomes (such as preventing pollution of a valued watershed). Indeed, the impressive research efforts of citizen scientists in the cases we studied often went unnoticed, barely registering in public debates about how to address these environmental challenges.

What hindered these projects from having more transformative effects? Most analysts would probably point to problems in the design of the projects—perhaps they did not aggregate their data effectively or lacked sufficient participation in all stages of the research. But having an impact with citizen science is not just a matter of good project design; it also depends on the features of the broader social context that influence both how citizen science is practiced and how it will be received. The effects of citizen science can be hard to predict because the social landscape is complex and full of contradictions. The practices that are now called citizen science have arisen out of multiple areas of social life, including education, science, government, industry, and social movements. Citizen science projects are pressured from many directions—changes in environmental policy, reductions in funding for scientific research, shifting cultural attitudes toward science, and criticism from industrial opponents. Doing environmental citizen science requires understanding and navigating this multidimensional social and policy environment.

Even thoughtfully designed projects may stumble over dilemmas that arise from the historical moment we inhabit. Throughout this book, we will explore questions concerning four general areas where citizen science faces recurring challenges.

1. Volunteering: Volunteering is always good for society, right? Maybe not. Social scientists are starting to recognize that some forms of volunteering are corrosive to important social values, such as inclusion, equity, and social justice. Reductions in public science funding have spurred the use of voluntary data gathering as a cost-cutting measure. Thus citizen science projects may inadvertently promote the idea that environmental problems can be adequately addressed on a voluntary basis—when, in many cases, law and policy are required. Furthermore, rates of volunteering vary widely across demographic groups and geographic regions, which can have problematic effects on the knowledge that citizen science generates. Citizen scientists who recognize the value and necessity of volunteerism must also grapple with these troubling patterns.

2. Taking a stand: Environmental movements have brought about crucial changes in how societies use natural resources, deal with waste, and relate to nonhuman species. Activists in these social movements are among the most important innovators in citizen science. However, they face a problem: science is typically perceived as tainted or biased if it is associated with a political cause or social movement, and opponents accuse them of "politicizing" science. Thus a crucial dilemma for environmental citizen scientists is how to establish the credibility of their scientific claims while maintaining their commitment to bringing about change.

3. Contextualizing data: Environmental citizen science expands scientific literacy and can lend scientific strength to environmental arguments. When environmental decision-making focuses on

technical information—as in risk-benefit assessments—citizen science can provide a way for ordinary people to participate. However, environmental issues involve values, emotions, social inequality, history, and aesthetics, which are not easily reduced to simple, rational measurements. Focusing on collecting scientific data can diminish the transformative power of citizen science if it inhibits discussions of power relations in society and marginalizes social and ethical concerns. For this reason, citizen scientists must carefully consider how to contextualize the data they collect.

4. Shifting scales: The phrase "Think globally, act locally" is widely applied to environmental issues, but it conceals a serious dilemma for citizen scientists. Community-based science can be effective at revealing localized impacts of polluting industries, providing fine-meshed and intimate data that experts often overlook. However, gathering local knowledge can sometimes imply personal responsibility for managing risks. When those responsible for the pollution are located in distant places and are governed by nonlocal authorities, scaling up is essential. For citizen science to have a role in solving environmental problems, participants must have a strategy for making change at the systemic level while remaining attentive to the local knowledge that otherwise would be ignored.

This book focuses on these dilemmas across several diverse examples of environmental citizen science. By unpacking the politics of citizen science, we aim to help people negotiate a complex political landscape and choose paths that bring about social change and environmental sustainability. We hope that by making the challenges and dilemmas explicit, we will help future citizen scientists make informed and thoughtful decisions about their strategies and methods.

What Is Environmental Citizen Science?

The idea of "citizen science" rapidly gained credibility and visibility among professional and academic scientists in the 2000s, and it is an increasingly global phenomenon. U.S. policy makers, for instance, started to use the idea of citizen science in the 2010s, arguing that science "Of the People, by the People, for the People" would enhance "data democracy."[5] Explaining the policy importance of citizen science, the White House Office of Science and Technology Policy observed that "only a small fraction of Americans are formally trained as 'scientists.' But that doesn't mean that only a small fraction of Americans can participate in scientific discovery and innovation. Citizen science and crowdsourcing are approaches that educate, engage, and empower the public to apply their curiosity and talents to a wide range of real-world problems."[6]

Under the Obama administration, the U.S. government took a number of actions that increased awareness of the concept, such as creating the website citizenscience.gov, which was "designed to accelerate the use of crowdsourcing and citizen science across the U.S. government." Congress passed several pieces of legislation that encouraged federal agencies' use of citizen science, including the 2013 America COMPETES Act and the 2017 American Innovation and Competitiveness Act. The vision of citizen science in these policies emphasized both innovation and transparency. For example, the Open Government National Action Plan of 2013 claimed that citizen science and crowdsourcing were "open innovation methods" that could "harness the ingenuity of the public."[7]

Similar actions to promote citizen science were taken in Europe. The European Union launched the Global Citizens' Observatory for Environmental Change in 2008, and in 2012, the European Commission launched SOCIENTIZE, a project to explore the potential of citizen science in Europe.[8] Horizon 2020, a large research funding scheme worth €80 billion (about $90 billion) over seven years (2014–2020) highlighted the importance of participation by citizens in science. It funded the ambitious "Doing It Together Science" program, which supports participatory projects in biodesign and environmental monitoring across Europe.[9]

Citizen science has been institutionalized with the establishment of the Citizen Science Association in the United States

Citizen science

(2012), the European Citizen Science Association (2013), and the Citizen Science Network Australia (2014). There is now an open-access scientific journal dedicated to citizen science, a *Citizen Science Day* that raises awareness about citizen science, and an online clearinghouse of citizen science opportunities called SciStarter.

Citizen science has gained traction in policy and research fields, but its definition has not been clear-cut, and the idea can be hard to define. In the natural sciences, the phrase was first used in the 1990s by Rick Bonney at the Cornell Lab of Ornithology, initially to describe the contributions of bird-watchers to the science of ornithology and later in several influential articles and books with collaborators in other natural sciences.[10] Bonney and his colleagues have used the term "to describe projects for which volunteers collect data for use in organized scientific research."[11] Following this tradition, natural scientists tend to conceptualize citizen science as research that employs volunteers as assistants or engages in crowd sourcing to acquire data.[12] But there are other uses of the term as well. In a 1995 book titled *Citizen Science*, sociologist Alan Irwin described how environmental contamination was forcing regular people to learn about scientific issues and how, in turn, they came to contribute to scientific debates.[13] Social scientists, following this line of analysis, have often used the term *citizen science* to describe the practice of science by social movement activists or to discuss efforts to democratize scientific inquiry. For example, activists may do citizen science for the improvement of health and the environment—often in cases of environmental racism or other forms of injustice.[14]

In 2014, the term was added to the *Oxford English Dictionary* (*OED*), defined as "scientific work undertaken by members of the general public, often in collaboration with or under the direction of professional scientists and scientific institutions." By this definition, citizen science is actually nothing new. Popular engagement in environmental science has a long history. Amateur naturalists and experimenters were considered respectable members of the scientific community until the nineteenth century. The subsequent professionalization of the sciences and the growth of formal academic institutions marginalized the amateurs, creating a gap between scientists and "lay people."[15]

Despite the professionalization of science, throughout the twentieth century, people without advanced science degrees participated in the process of scientific inquiry through "volunteer monitoring" or "citizen's watch" projects. The U.S. National Weather Service created the Cooperative Observer Program (COOP) in 1890, which recruited volunteers to take daily meteorological data.[16] The National Audubon Society's Christmas Bird Survey in the United States and Canada is also more than one hundred years old. Other initiatives, such as the North American Breeding Bird Survey or the Butterfly Monitoring Scheme based in the United Kingdom, began in the 1960s and 1970s. The U.S. lilacs and honeysuckles phenology observation was initially started by the U.S. Department of Agriculture in the 1970s and since has been revitalized and maintained by other institutions.[17] In Japan in the 1960s, citizens worked with scientists to observe mutations of spiderworts near nuclear reactors; in later decades, Japanese citizens helped monitor dioxin emissions from waste incinerators.[18] The U.S. Clean Water Act in 1972 generated strong interest in volunteer watershed monitoring across the country, aimed at understanding whether laws and cleanup efforts were having the desired effect on water quality.[19]

The *OED*'s definition of *citizen science* leaves open questions about what constitutes "scientific work." Many things that people do on an everyday basis could be considered scientific work, such as growing yeast to raise bread, thinking about statistics presented in the media, exchanging information with peers about health experiences, and making hypotheses about the sources of pollution they observe. At the same time, the definition of *science* has historically excluded different epistemologies such as knowledge of indigenous peoples and their ways of understanding and interpreting the world. Furthermore, some projects that have been labeled citizen science do not fully engage participants in what we would consider scientific inquiry. For instance, some citizen science of has been "gamified," as in the example of Foldit, an online video game in which participants solve protein-folding puzzles, which helps university researchers understand how living beings create the primary structure of proteins.[20] In contrast, others, such as a citizen science project on environmental health in Europe headed by science and technology studies (STS) scholar Barbara L. Allen, have engaged community members much more deeply and at various stages of research, from agenda setting and survey designs to data analysis, which Allen calls "strongly participatory science."[21] Therefore, the degree of participation by community members varies quite significantly even as projects take "participation" as a hallmark.

Additionally, there is a question of who constitutes the "general public" in the *OED*'s definition. Aren't professional scientists members of the public too? Does the use of the phrase *citizen science* suggest that professional scientists are somehow not citizens? The phrase is also problematic because it implies that participation depends on one's citizenship status, which can marginalize indigenous and immigrant communities. Participatory research could involve collaboration across national borders and with people of multiple political communities, straining the concept of "citizen" science.

In this book, we have chosen to focus on projects in which people who are not professional scientists take part in structured, collective efforts to investigate some aspect of their environment. Despite the problems we identified in the previous paragraphs, we use the umbrella term *citizen science* to refer to these practices, since the term is now widely in use. However, we also use a wide range of descriptive terms such as *grassroots science*, *collaborative research*, and *participatory environmental monitoring*, aiming to clarify the kinds of scientific work being done and the style of participation that it involves. We also draw on the terminology used in a variety of other research practices. These include participatory action research and community-based participatory research (CBPR), two approaches to research in which professionals collaborate with community partners to investigate and take action on a shared matter of concern.[22] A related concept is popular epidemiology, which refers to the ways that lay people collectively detect and make sense of the experience of environmental illness.[23] In geography, participatory mapping enables researchers and communities to represent places with features that are important to the people who live there.[24] Terms such as *DIY science* and *critical making* are often used to describe movements that develop alternative tools for scientific research, such as Genspace, a community biotech lab; DIYbio.org; X-Clinic; and Sensemakers.[25] Each of these areas of work provides us with terminology to describe participatory scientific work.

To be sure, we are not the first to highlight the diversity of citizen science practices. One common approach to writing about citizen science is to offer a typology. Many typologies focus on the structure of participation, distinguishing among the myriad forms that relationships between laypeople and professional scientists can take.[26] Others create types based on project goals, such as action (to support civic agendas), conservation (to support environmental stewardship), investigation (to support scientific research), and education (to support formal and informal learning).[27] Typologies can be useful, but they often obscure the significant commonalities and interconnections among different "types" of citizen science and the ways that projects can change over time. We have also observed that typologies can lead people to conclude that they practice the superior type of citizen science—hindering exchanges of ideas with those taking different approaches. We have therefore chosen not to sort environmental citizen science into types; instead, we recognize and examine the messiness and diversity of this broad category of participatory research practices. We hope that the juxtapositions of examples that we make in the chapters that follow will lead readers to fresh insights about their own research practices.

As the concept of citizen science travels around the globe and into public policy, its definition may continue to shift to encompass new practices. At the same time, we are alert to the possibility that as citizen science gains stature in the worlds of policy and

science funding, some practices—such as research by environmental activists—may not be recognized as citizen science by
professionals.

This book focuses on *environmental* citizen science projects,
which engage people who are not professional scientists in studies of natural resources, ecosystems, biodiversity, and pollution.
This leaves out a wide swath of the world of citizen science, which
addresses a vast array of topics. For example, in the Milky Way
project, volunteers look at images of the Milky Way galaxy and
outline features that are of interest to astronomers.[28] The British
Gut Project organized by King's College London crowd sources
samples from citizens with the objective of creating a database of
gut microbes.[29] While fascinating, projects like these are outside
the scope of our analysis. Nevertheless, we hope that the lessons we
draw will be useful for people who are working in other issue areas.

What Is It Good For?

There has been an outpouring of academic research and writing
about citizen science, both by scientists who are incorporating citizen science into their research and by experts on science policy,
education, and civic engagement. This research sheds light on the
wide-ranging approaches to participatory research and the contributions that nonprofessional researchers have made to knowledge
and society. Most scholars see citizen science in a positive light,
apart from some concerns about data quality that we discuss in
what follows. However, authors are not all on the same page about
why citizen science matters.

When professional scientists endorse citizen science, it is typically because it helps them collect and analyze very large quantities
of data. Discussions about citizen science in the field of ecology frequently draw attention to the expense of research, as in one study
on the prediction of emerging plant diseases, which indicates that
"the economic cost of collecting sufficient data . . . is often prohib-
itive but may be offset by involving volunteer citizen scientists in
the data collection process."[30] Citizen science enables important

research on pressing issues relating to climate change, disease, and species extinction, among others, and can lead to scholarly publication of results. For example, Earthwatch, an international environmental charity, has sponsored citizen science projects that have yielded hundreds of peer-reviewed publications.[31]

The monetary value of free labor provided by citizen scientists is estimated to be large, even though there may be high costs of preparation, recruitment, training, and retention of participants. For instance, biodiversity researchers who analyzed 388 citizen science projects estimated that volunteers spent an average of twenty-one to twenty-four hours per person annually collecting data. Each volunteer's "in-kind contribution" of work was worth $1,900, totaling $667 million to $2.5 billion annually. The researchers observed, "The time donated by the large number of engaged volunteers (approximately $0.7–2.5 billion annually) is equivalent to 11–42% of the annual U.S. National Science Foundation budget."[32]

The relatively lower costs of data produced by citizen scientists have strong appeal, but will the scientific contributions be worthwhile? Successful publication of citizen science–derived data in peer-reviewed academic journals is a challenge.[33] E. J. Theobald and colleagues surveyed more than three hundred biodiversity citizen science projects and found that only 12 percent of them had resulted in peer-reviewed scientific publications. This was despite the fact that 97 percent of them explicitly aimed to advance scientific knowledge.[34] The reasons for the low rate of successful academic publication are complex. Scientists may not be aware of citizen science–generated data. Quality assurance and quality control of citizen science data have also been criticized. For instance, volunteers might misidentify different varieties of plants and animals or be more interested in and hence more likely to report on rare species rather than common ones. The low rate of journal publication may also be related to the perception of scientists. Many seem to prefer data from traditional academic institutions and assume that citizen-generated data, particularly generated by people with little education, is necessarily of suboptimal quality.[35]

Scientists may also worry that if their field becomes strongly associated with public participation, they may be perceived as less valuable, shoestring-budget science.

In this context, some researchers are now pursuing "science of citizen science," investigating ways to reduce the barriers to using such data for scientific research.[36] Some are interested in the traits of citizen science projects that tend to produce academic publication results (larger scale, longer project duration, affiliation with academic institutions).[37] There is also increasing attention to ways to enhance data quality, such as training by professional scientists, pre- and posttests, competency tests of citizen data collectors, identification and exclusion of data from unreliable contributors, and aggregation of citizen data.[38] Scientists' perceptions of citizen science may be changing as well. For instance, influential and high-impact journals such as *Conservation Biology* have had special sections featuring citizen science in recent years.

At the same time, the scale of volunteer contributions to the scientific profession raises questions about the science funding regime and the sharing of benefits arising from volunteer labor. Rebecca Lave characterizes such projects as using "unpaid labor" to advance the research agendas of professional scientists, highlighting one project that "is supported by volunteers who not only pay fees to participate, but also contribute an estimated $3 million per year in unpaid labor."[39] Moreover, Philip Mirowski points out that "citizen science is fuelled by the fact that the public sector is trying to get out of the science business," reducing public funding for scientific training and research. At a time when "support for universities has been squeezed, offices of science policy have been shuttered, and research has been willingly palmed off to private sponsors. . . . Policymakers have been especially enamoured with citizen science because it appears to promise the next stage of outsourcing and divestment of their own prior research portfolio—now by farming work out to the 'crowd.'"[40]

Yet professional scientists are not the only ones finding value in citizen science—and the value is not always monetary. For example, conservationists and education researchers contend that participation in citizen science increases scientific literacy and makes people more aware of environmental problems and ecological systems. Since the 1980s, low scientific literacy in the United States, particularly shown in the international comparison of test scores in science and math, prompted reforms in school curriculum as well as programs to encourage science learning in nonschool settings.[41] Volunteering in scientific projects—citizen science—emerged as one venue of improving the scientific literacy of the general public. For instance, the National Science Foundation (NSF) Public Participation in Scientific Research (PPSR) program, which encourages research involving public participation, has roots in the NSF's Informal Science Education program.[42]

Educational outcomes are the most frequently studied effect of citizen science projects, according to one meta-analysis of the literature.[43] Most studies, unsurprisingly, show that citizen scientists learn content knowledge about the subject they are studying, although if participants already have high levels of science education, such gains may not occur.[44] Some studies suggest that participants also gain skills related to the process of environmental monitoring as well as skills for communicating findings.[45] There is also research to suggest that participants gain an understanding of ecological processes, which may change their assumptions about environmental issues.[46]

Social scientists who are interested in political engagement and collective social behavior emphasize the ways that participation can help communities build capabilities to participate in conservation. For example, rural sociologist Rich Stedman and his collaborators found that watershed organizations in the U.S. state of Pennsylvania were "effective mechanisms for building local leadership, enhancing the skills of rural residents, and making valuable connections with other communities, facing similar water-resource and rural-development issues."[47] Likewise, sociologist Christine Overdevest and her colleagues found that experienced participants in a watershed monitoring project "were more active than their inexperienced counterparts in resource management-related behaviors, such as talking with and providing information to neighbors about resource issues, engaging in personal reading and research about resource issues, and attending public meetings to discuss issues." Their findings led them to speculate that similar programs could generate a "more engaged and connected citizenry," with the important caveat that participants in their study tended to be highly educated with high incomes.[48]

In the European context, citizen science recently has been connected to the concept of responsible research and innovation (RRI), an idea that was spurred by socially controversial technologies such as stem cell research and nanotechnology.[49] Research institutions and regulatory authorities began experimenting with practices such as deliberative conferences, participatory technology assessment, and science cafés—as well as citizen science. The idea of citizen involvement for the sake of RRI has two faces. The concern can be both with ethicality and accountability of scientific experts and with social acceptance of new technology by way of "stakeholder engagement."[50]

Some researchers suggest that citizen science is beneficial when confronting "wicked problems," which are not easily addressed by traditional discipline-based science done by professional scientists. "Wicked problems" refer to those issues that are complex and involving many stakeholders, such as climate change and conservation. Traditional science, the argument tends to go, is too siloed and myopic to effectively deal with these issues. Professional scientists are too captured by the rigidity of disciplines and requirements by the academic institutions. This prognosis has led to many calls for interdisciplinary research to create collaboration across areas and tackle real-world problems. Some have also indicated that citizen science has this virtue. Justin Dillon, Robert B. Stevenson, and Arjen E. J. Wals, for example, call for a more prominent role of citizen science in conservation: "To deal with these wicked issues, one needs to realize that citizens have or need to have agency; scientific knowledge includes other types of knowledge, for instance, indigenous knowledge and local knowledge;

and actions to improve a situation require social learning between the multiple stakeholders affected by an issue."[51]

Others contend that valuing alternative ways of knowing, such as indigenous and local knowledges, helps improve scientific understanding by diversifying perspectives.[52] Research in this vein emphasizes the positive effects of leveling status distinctions between experts and the lay public in participatory projects.[53] One example can be found at the Coweeta Hydrologic Laboratory in North Carolina, one of the most established research stations on forested watersheds in the United States since the 1930s. There, in 2010, geographer Nik Heynen and his collaborators started the Coweeta Listening Project, which aims to integrate the local knowledge held by people in the community with scientific research. They institutionalized dialogues with community organizations to identify areas of concern and challenge knowledge hierarchies. The project's website states this rationale for working in close collaboration with people in their region: "The co-production of ecological knowledge—which involves integrating scientific and non-scientific ways of knowing and bridging science and values—will enhance scientific research and make it more useful for addressing environmental concerns. In short, we hope to democratize science so that it is more effective and useful."[54]

Research focusing on environmental justice and social change tends to emphasize how citizen science can be oppositional and help marginalized communities challenge the powerful in society. Participatory research can enable ordinary people to critique expert claims—for example, by challenging denials of environmental health problems.[55] In this way, citizen science can be a tool to support social justice struggles. For instance, in the aftermath of the Deepwater Horizon oil spill in 2010, an environmental justice organization called the Louisiana Bucket Brigade tried to uncover the extent of the damages that might not have been captured by the government agencies and the responsible corporate actors. The group crowd sourced pollution observations and created the Oil Spill Crisis Response Map. The map covered a broader range of issues than the government agencies, allowing contributors to add data including odor, oil on water, properties damaged by oil, and impacts on wildlife and birds. The map also covered broader geographic areas than the government monitoring.[56] Out of the collaborative mapping response to the BP disaster arose Public Lab (PublicLab.org), a network of groups and individuals that create and share hardware and software virtually to address local environmental issues. STS scholar Sara Wylie and her collaborators describe Public Lab as challenging the traditional expert-citizen hierarchy, since citizens are involved in setting research questions, solving problems, and critiquing the claims made by professional scientists.[57]

Other perspectives on citizen science emphasize policy relevance as a virtue. Knowledge produced by citizen scientists can support arguments for changing public policy and regulatory strategies. In the book *Street Science*, urban planner Jason Corburn describes a case in which the Environmental Protection Agency (EPA) initiated a study to model cumulative exposure to toxic pollutants in the Greenpoint/Williamsburg neighborhood of Brooklyn, New York. Although the project did not begin with a plan for participatory research, a first step was to hold community meetings in order to consult with residents. Through those meetings, the EPA learned that the agency was overlooking a major dietary risk: the consumption of fish caught from the East River. Initially skeptical of the community's concerns about fish consumption, the EPA worked with a community organization, the Watchperson Project, to learn more about this potential risk. EPA helped the Watchperson Project gather data from anglers and ultimately used those findings in its calculations of lifetime cancer risks and recommendations about fish consumption.[58]

Finally, if some discussions of citizen science emphasize interventions through challenging authority or promoting policy change, others focus on the virtue of catching polluters in the act of breaking environmental rules.[59] There are numerous examples of citizen scientists, particularly in the case of air quality monitoring, drawing attention to regulatory violations.[60] Yet there have not been systematic studies of the effectiveness of citizen science for this purpose. Surveys of organizations that do volunteer water monitoring suggest that environmental policing is not a major aim of these projects compared to the more prevalent goal of raising environmental awareness.[61] However, in a study of citizen scientists who are tracking the impacts of the natural gas industry on watersheds in Pennsylvania, Abby Kinchy and her collaborators found that a "policing logic" guides the work of some monitoring groups.[62] The volunteers test water samples regularly and report any signs of pollution or disturbance (such as stream-bank erosion) to regulatory authorities. Some refer to themselves as extra "eyes and ears" for state regulators. Others, using sensors to accurately track changes in water quality on a continuous basis, hope that their data will stand up to scrutiny in a court of law. It remains unclear, however, how much these efforts have actually improved the enforcement of environmental law.

Clearly, there are many different perspectives on what citizen science is good for. That there are multiple possible virtues of citizen science suggests that many people will have reasons to participate and support it. Yet a virtue from one perspective might be a drawback from others, or emphasis on one might mean deemphasizing other issues. For instance, in their review of the literature on conservation biology and citizen science, conservation biologist Mark Chandler and colleagues argue that "citizen science projects are often challenged to simultaneously maximize different goals, which appear to create trade-offs between producing both scientific publications and management action."[63]

While citizen science can be admired for increasing scientific literacy among participants, critics might argue that projects that focus on learning scientific and technical content are less likely to stress the sociopolitical roots of environmental problems. Likewise, the virtue of collecting large amounts of data to support professional scientific research stands in contrast to the virtue of producing "oppositional" knowledge, or challenging expert-public distinctions. Which virtues should citizen scientists prioritize? This is ultimately a question of values, and a goal of this book is to encourage citizen scientists to pay careful attention to the virtues they pursue and values that guide their orientation in citizen science projects.

Our Methods

Our approach to this study employs C. Wright Mills's concept of sociological imagination, which refers to a quality of mind that grasps "the interplay of individuals and society, of biography and history, of self and world."We probe the relationships between particular, unique examples of citizen science and the broader history and society of which they are a part. To understand the interplay between on-the-ground practices and social structures, we combine analyses of secondary materials (writings on citizen science by other scholars) and our own data on specific citizen science projects that we have investigated over several years. The use of secondary materials is necessary not only to show that the idea of citizen science is gaining traction but also to identify patterns across diverse cases. Although we seek to identify trends and patterns, we note the importance of contexts and historical contingencies. We combine a broad overview with more detailed case studies, presented in chapters 3, 4, and 5. The case study chapters draw on our own extended fieldwork on the issues of fracking in the United States (Kinchy), radiation monitoring in Japan (Kimura), and monitoring of genetically modified organisms (GMOs) in Mexico (Kinchy) and in Japan (Kimura), and more details about the methods of each study are provided in the endnotes of those chapters.

We have pointed out that citizen science practitioners often implicitly bring in values that lead them to emphasize some virtues and utilities of citizen science over others. As we try to clarify our methodology, we believe it is important that we make clear assumptions and our own values. We start from the understanding that science is implicated in culture, society, and politics and that it is necessary to consider such social forces as essential components of scientific inquiry. We do not assume that any science is neutral or free of social relations[64]—that is, while a nonprofessional scientist may be different from someone with an advanced degree and a research position, both are social beings, and their efforts to understand the natural world are shaped in some ways by their cultures, political systems, economic conditions, and other aspects of society. Furthermore, we do not assume that citizen science is best when it approximates the practices of professional science; instead, we explore the great variety of practices of environmental inquiry, recognizing that professional science has its own limitations.

As C. Wright Mills said, anyone, not just professional sociologists, can develop a sociological imagination. Thus one of the aims of this book is to foster a sociological imagination about citizen science and the ways that science relates to environmental politics. This is necessary—even more so today than at the publication of these lines of *The Sociological Imagination* in 1959: "It is not only the skills of reason that [ordinary people] need—although their struggles to acquire these often exhaust their limited moral energy. What they need, and what they feel they need, is a quality of mind that will help them to use information and to develop reason in order to achieve lucid summations of what is going on in the world and of what may be happening within themselves."[65] We hope to guide readers to develop the "quality of mind" that Mills describes when they think about their own practices of citizen science and how they are gathering and using information. Throughout this book, we seek to identify and build reflexive moments that foster ongoing and creative engagements by practitioners and theorists of citizen science.

Plan of the Book

In the next chapter, we make the case that citizen science is a political act even when participants try to avoid taking a controversial stance. We are interested in the ways that citizen science not only produces scientific answers but also challenges social inequalities and creates a healthy and flourishing environment for all. Here we introduce the ideas about the politics of knowledge, environmental justice, and citizenship that form the basis of our sociological analysis of citizen science. Then we examine the forces behind the increasing public policy support for citizen science and introduce the concepts and frameworks for exploring the four dilemmas summarized previously.

In the subsequent three chapters, we present case studies that enable us to more deeply probe these political questions and dilemmas around citizen science. One aim of these chapters is to show that citizen science happens in and is shaped by political contexts that go beyond the local issue or controversy. We highlight the problems that citizen scientists face when deciding how to carry out their projects because of the complex and contradictory institutional environments they inhabit.

Admittedly, the examples of citizen science that we have in our empirical chapters are mostly focused on monitoring contaminants of water, food, air, and soil. Some readers may wonder about other kinds of environmental citizen science, such as studies of endangered species and wildlife habitat. We recognize this limitation and the possibility that dilemmas other than those we identify in this book may arise in different kinds of participatory research. Our central questions, however, are relevant to many citizen science projects. For example, How can participatory research help people stand up to powerful opponents, gain voice in public decision-making and agenda setting, and address inequitable distributions of benefits and harms?

In chapter 3, we examine citizen science in the context of the shale oil and gas boom in North America, looking closely at water monitoring in Pennsylvania and New York. Since the start of natural gas development (fracking) in the Marcellus Shale around 2007, dozens of volunteer water monitoring groups have initiated efforts to track the impacts on local streams. Their work builds on a history of volunteer watershed monitoring that traces back to the early twentieth century but has new significance in a context where state and federal agencies appear unable or unwilling to rigorously investigate the impacts of fracking in rural communities. Other citizen science efforts, such as air quality monitoring and observation of transportation infrastructure and industry practices, illustrate the diversity of practices and the range of possible strategies and outcomes. In this chapter, we begin to more deeply probe the issues that arise in relation to volunteering, taking a stand, contextualizing data, and shifting scales. For example, we observe that participatory projects often seek to substantiate residents' concerns about pollution by taking quantitative measurements that can be compared to regulatory standards. Yet this can lead to a "data treadmill" in which considerable energy and time are invested in defending the accuracy and relevance of the measurements while the broader set of concerns about how oil and gas development is affecting communities and climate is sidelined.

In chapter 4, we focus on the case of radiation monitoring after the Fukushima nuclear disaster in Japan. Frustrated by the lack of information from authorities after the accident, lay citizens took matters into their own hands, establishing projects to examine food contamination and other areas of potential radioactivity, such as ocean water and soil. Looking in detail at the case of food testing, we see several difficult dilemmas. A strength of citizen science is that it can produce finely grained data that can be used by individuals for self-monitoring and protection. Investing in tools for radiation monitoring might be one of the very few means that regular people have to reduce their radiation exposure. But simultaneously, monitoring at the individual scale can shift the work of protecting people away from the government (and the industry) to affected individuals. We investigate how this situation arose along with other problems related to citizen science after Fukushima.

Chapter 5 crosses several continents where activists sampled food and grains moving through the global food system for traces of genetically engineered (GE) crops. From the scandal of food contaminated with GE corn that was not approved for human consumption, to the controversy over unapproved GE corn in Mexico, to the problem of GE rapeseed growing wild in Japan, there are many examples where citizen scientists have illuminated the difficulty of controlling where living GE plant species go once they are released into the environment. Notably, many activist-led monitoring projects aim to highlight socioeconomic challenges, such as the effects of globalization in the food system and the dominance of multinational corporations. The chapter also considers examples of citizen science that the organizers portray as neutral and nonpartisan but which nevertheless advance certain values and commitments. These examples enable us to consider the ways that participatory studies can facilitate (or deter) multidimensional analysis of social-natural systems.

In the concluding chapter, we draw lessons from the previous chapters to suggest how citizen scientists can negotiate a complex political terrain and choose paths that bring about social change and environmental sustainability. There is no "one-size-fits-all" method of citizen science that will bring about desired social changes. Instead, what we argue is that citizen science projects require reflection and analysis not just about research design and data collection tools but also about strategies for negotiating the issues that are generated by the broader political environment. The key takeaway of this chapter is that citizen science is a political act, requiring ethical reflection, strategic planning, and a sociological imagination.

2

How Is Environmental
Citizen Science Political?

New York City gardeners are using citizen science to defend against threats that their plots will be sold to private developers. In 2010, community gardens on public land were at risk of losing the protections that had preserved open spaces for gardening. Government officials, policy makers, gardeners, and activists believed that having more data on the city's community gardens would strengthen advocacy to preserve them. To meet this need, gardeners in a nonprofit organization called Farming Concrete used weighing scales to measure garden harvests, calculating their economic and nutritional benefits. A 2011 report by Farming Concrete concluded that New York City community gardeners were able to grow some crops more efficiently (in terms of space) than conventional farmers, with yields that exceeded the country's average.[1] Farming Concrete subsequently branched out, creating a tool kit that is now used nationwide. With input from community gardeners, the methods have expanded beyond weight measures to include environmental, social, health, and economic data.

The political aim of the project is explicit: to protect the land used for urban community gardening.[2] Yet public materials about Farming Concrete do not address the stickier political dimensions of community gardening, such as questions of fairness or inequality in access to urban green space or the role that community gardens sometimes play in the gentrification of urban neighborhoods.

This is not unusual—citizen science projects cannot be expected to address every dimension of the issues they face—but it does prompt reflection on how participatory research fits into the broader political terrain. In this chapter, we begin to probe the politics of citizen science and the ways that participatory environmental research can support struggles for social change.

Voice, Knowledge, and Power

Citizen science provides scientific observations and data, but it can do more. It can help make science more diverse and inclusive, provide a voice to the marginalized, and challenge power inequalities. There is no simple design of citizen science that guarantees these transformative outcomes. As we suggested in chapter 1, an array of dilemmas can make it difficult to pursue these aims. As a result, there is wide variation in how citizen science projects pursue or attain these goals. Before addressing the dilemmas, let's consider the transformative potential of participatory environmental research.

First, citizen science can help *give voice to marginalized communities and create a platform for political participation*. Participating in scientific inquiry can be a means to gain standing in decision-making processes and can help people make policy proposals that rectify long-standing injustices. For example, Farming Concrete's measuring efforts helped urban gardeners voice their opposition to selling the garden plots to private developers.

This possibility invites critical reflection on the *citizen* part of *citizen science*. While it seems to be a straightforward idea, this term can be interpreted in many different ways. For instance, two classic traditions in political theory, liberalism and civic republicanism, assume that citizens are members of a political community with the same rights and responsibilities. Civic republicanism further emphasizes the practices of participation in the political sphere, where citizens collectively decide about community affairs and share a common life with other citizens. These views of citizenship tend to consider citizens as homogeneous, with universal rights and responsibilities and a shared culture. Citizen science that is pursued in the spirit of these traditions tends to ignore the ways that racism, sexism, and other forms of oppression determine who can speak and be heard in public life.[3]

Our perspective on citizenship calls for considering histories of oppression and marginalization that position people differently in the political community. In this view, a just society would address structural inequalities along the lines of gender, race, ethnicity, sexuality, ability, and so on because a difference-blind approach to law and policy privileges the already powerful. Feminist philosopher Iris Marion Young contends that "the inclusion and participation of everyone . . . sometimes requires the articulation of special rights that attend to group differences in order to undermine oppression and disadvantage."[4] Citizen science pursued in this spirit could help open up space for people whose perspectives on the environment, natural resources, and energy are not often heard. For instance, as we will discuss in chapter 5, residents in Japan have been collaboratively investigating the ongoing presence of radioactive material in foods, years after the Fukushima nuclear disaster. Women, particularly mothers, are often denigrated as too emotional and therefore harshly critiqued in Japanese policy debates for their radiation concerns. Participatory research that emphasizes the special value of women's knowledge and lived experiences may help undermine the disadvantages that Japanese women experience as a group.

Having a voice in public life is no longer limited to local or national debates. Contrary to nation-state-based ideas of citizenship, the field of citizenship studies is increasingly pointing to the layered, shifting, and contested nature of citizenship in the contemporary world, defining citizenship more broadly as an "institution mediating rights between the subjects of politics and the polity to which these subjects belong."[5] In this definition, a "polity" is not necessarily limited to the country that issues a passport for the subject, and "belonging" is not necessarily limited to long-term physical residence. From this perspective, citizenship is not a static set of legally codified rights and duties of passport-holding people, but it is something that is enacted and negotiated in different contexts with different polities. Rather than static and singular, citizenship rights are increasingly understood to be negotiated and performed in diverse ways.

These ideas encourage us to explore how citizen science expands possibilities for people to perform citizenship—to have a public voice—in relation to different polities. For example, citizen science projects may help undocumented immigrant farm workers in the United States have a say in pesticide regulations although they are not naturalized Americans. We believe that citizen science's potential can be maximized when it helps historically marginalized groups of people—including people who might be considered "noncitizens"—make arguments for their participation in decision-making and for their rights and respect.

This leads to our second observation about the transformative potential of citizen science: it can *produce knowledge from vantage points historically left out of science.* Academic science as a social institution has been highly elitist, Western-centric, and male-dominated throughout most of its history. Academic institutions and research organizations have historically been populated by well-to-do men of Western European origin, and people of color, the poor, and women are underrepresented in positions of scientific authority. While science is diversifying, this history continues to have an effect on long-term research agendas and ideas about what kinds of questions are interesting and important.[6]

Research funding priorities tend to reflect the concerns and interests of the socially powerful, so there is often a "mismatch between the knowledge that science generates and the knowledge society needs."[7] In issues ranging from uranium mining to bee colony collapse, concerned citizens have called for more research into the harms produced by powerful industries. Furthermore, funding for scientific research increasingly comes from the private sector, steering researchers toward projects that can result in intellectual property and economic benefits for commercial actors.[8]

Citizen science can be socially transformative when it responds to this problem and offers alternative knowledge and interpretations of bodies, environments, and social worlds. Environmental movements provide great examples of this potential: many have had positive impacts on scientific discovery, bringing new problems to light and forging "alternative pathways" in research and development.[9] For example, air pollution monitoring by communities bordering petroleum refineries has challenged missing information in official scientific accounts. In Norco, Louisiana, residents of a predominantly African American neighborhood bordering the Shell Corporation refinery experienced numerous health ailments that they understood to be caused by toxic releases from the facility. However, the air quality data gathered by the Louisiana Department of Environmental Quality and U.S. Environmental Protection Agency (EPA) only measured the average concentration of toxic chemicals over time and did not capture sudden spikes during and immediately after their release from the refinery. The Louisiana Bucket Brigade, an advocacy group, trained local residents to deploy a simple air-sampling method that enabled them to send plastic bags of suspected air pollution to a laboratory for chemical concentration analysis. Their data supported grassroots activism against the plant and convinced government regulators to investigate further.[10] In this case, citizen science not only facilitated or extended the work of professional scientists but offered a challenge to official accounts by taking seriously the experiences and knowledge of a poor and marginalized community.

Of course, citizen science itself can create silences or leave out marginalized social groups. The ideas of "local" people and "communities" often imply clear boundaries and internal homogeneity, but defining the community can be contentious, and there are important differences in perspective and experience within any locality or social group.[11] Furthermore, some people may be excluded if a citizen science project requires a significant commitment of resources and effort. Women, who tend to shoulder more caring responsibilities and domestic work, may have less time to participate in citizen science. This could shape the research agenda of a citizen science project in ways that leave out the perspectives of women. Similar problems may arise for the working poor, people with disabilities, and other groups that face greater obstacles to participation. For instance, plant ecologist Heidi Ballard has written on a participatory research project for natural resource management. She points out how a "local community" is not homogenous even as it relates to the same forest resource and how immigrant status may hinder some groups from participating in research and resource management. She writes,

> The legal status of immigrant forest workers plays a huge role in the ease with which researchers, managers and non-governmental organizations might promote inclusion in natural resource management. Many workers in the Pacific Northwest are *sin papeles* (without papers), and contacting workers through contractors proves difficult. In the case of pineros, the fact that many forest workers were related to their employers inhibited their willingness to participate in collaborations that might have jeopardized their employers, supervisors and themselves. The undocumented status of many floral greens harvesters in western Washington also created obstacles to participatory research on salal [floral greens]. For example, introduction to the community via federal agencies, which might have proven helpful in other circumstances, was relatively unproductive because harvesters were justifiably wary of any possible connection to the Immigration and Customs Enforcement.[12]

Environmental sociologist Jill Harrison, who has researched pesticide monitoring projects, has likewise observed that the people most affected by pesticide drift are often not U.S. citizens, and because of their immigration status (particularly if they are undocumented), they lack standing to demand health protections.[13]

Citizen science is therefore not necessarily and not automatically inclusive of marginalized perspectives, nor does it guarantee a level playing field for all participants. However, it can be transformative when these limitations are addressed in a thoughtful manner. Citizen science practitioners can purposefully consider the silence and invisibility of certain perspectives on the topics they are investigating.

Giving voice to the marginalized and diversifying knowledge production give rise to the third transformative capacity of citizen science: *building political power to challenge environmental inequality*. Exposures to pollution and environmental destruction are not evenly shared; in the United States, African Americans, Latinos, and Native Americans live in disproportionately polluted environments.[14] In citizen science projects that are motivated by environmental justice, participants grapple with how inequality makes some social groups—such as the Norco community introduced earlier—more vulnerable to toxic wastes, hazardous industries, and environmental illnesses. Participatory, collaborative research can enable people to make informed decisions, advocate for their rights, and challenge developments that disproportionately harm historically oppressed people.

Environmental justice activists tend to use citizen science to increase the credibility of their claims and shift media and policy discourses, with the goal of changing the political and economic structures of environmental contamination. While experts, polluting industries, and the general public often disbelieve claims of environmental illness, citizen science can challenge environmental injustice by providing data that can be used in lawsuits, bureaucratic proceedings, and media representations. Of course, credentialed experts, opponents, and regulators may cast aspersion on citizen science data as unprofessional or biased. But having data is useful when environmental governance privileges science as the most legitimate avenue to register concerns by the public. In these ways, citizen science can be empowering.

Our use of the word *empowering* here could easily be misconstrued, since citizen science projects that use volunteers to monitor the environment frequently employ the idea of empowerment. For instance, Greg Newman and colleagues, developers of CitSci .org, define citizen science as focusing "on data collection by volunteers across broad geographic regions with the aim to inform and empower citizens and benefit scientists, land managers, and decision makers."[15] Certainly, contributing to knowledge production can produce a sense of self-efficacy and social contribution for many volunteers, but when we talk about building power through citizen science, we are concerned with power relations among individuals and institutions in society. Empowerment, from our perspective, means having the capacity to implement changes to the systemic and structural sources of environmental problems. It is important to distinguish this specific meaning from the more generic and loose definition of the term.

Sociologist Nina Eliasoph points out that many volunteer organizations today employ "empowerment talk," a mantra "heard around the world, wherever officials try to cultivate grassroots community empowerment, from the top down." Empowerment talk calls for civic participation, innovation, diversity, inspiration, looking forward, working at the community level, transparency, and proceeding without reliance on distant experts.[16] Volunteer projects in this vein emphasize self-improvement and self-reliance rather than collective action for social change. Eliasoph observed that in volunteer projects targeting at-risk youth, a "relentless, unreflective focus on future hopeful potential" frequently cut off discussions about the historical reasons for the problems these communities and youth were facing, such as structural racism and poverty.[17] In this way, the idea of empowerment through volunteering often dulls the edge of work to address systemic injustices.

Eliasoph's research raises an uncomfortable question about volunteer-based citizen science projects: Could they lead people to channel their concerns *away* from confronting the underlying sources of their troubles? For example, volunteers may learn how to protect themselves from pollution or make small-scale environmental improvements rather than challenging those in power. They may be led to believe that as more information about environmental problems becomes available through citizen science, polluters and policy makers will "do the right thing." Neither of these common situations is conducive to building the power needed to exert influence over the industries and political institutions that produce environmental harms.

When we focus on voice, knowledge, and power, we can see that citizen science is not simply about data collection and analysis but can be a tool for social transformation. It is also clear that citizen science cannot be divorced from historical patterns and social structures. Doing citizen science unsettles existing patterns of knowledge production that have created certain blind spots and silences. Citizen science can be a tool in environmental justice struggles to challenge entrenched power structures. These transformative potentials of citizen science are not simply obtainable by following a set recipe; rather, the potential of citizen science is always a result of complicated negotiations with an array of dilemmas, to which we now turn.

Volunteering to Serve the Public Good

Citizen science can be considered an instance of volunteerism—an altruistic public service by regular citizens for the creation of scientific knowledge, a public good. Yet the rise in participatory data gathering is driven, in part, by broader political trends, including reduced public funding for environmental monitoring and science, as well as a particular kind of discourse about volunteering and citizenship.

Many people would consider volunteering to be an act of good citizenship. Volunteer environmental monitoring, for instance, can reduce resource demands on government agencies and help them achieve their mission. Volunteerism can improve the vibrancy of civil society, bringing more people into public-spirited projects. Sustained participation in volunteer groups, like watershed monitoring teams, enables people to gain skills, knowledge, relationships, and mutual trust that are necessary for participation in democratic life. Volunteering can lead to political action and advocacy for change at local and national scales.

Many citizen science projects have been shown to have these virtues, but social scientists observe that certain forms of volunteering are more effective than others at bringing positive changes to social and political life.[18] Volunteering today is often related to ideologies of small government, market-based problem solving, and the rollback of regulations, social programs, and research funding. Since the 1970s, the United States and many other countries have experienced a shift in political philosophy and public policy that many scholars refer to as neoliberalization (referring to the renewed embrace of early theories of the "liberal," or free market, economy). Neoliberal policies emphasize "small" government, private funding for formerly public services and research, and individual responsibility for health and welfare. An increased emphasis on volunteerism is consistent with this approach to governance. As neoliberal ideas become taken for granted as "the way things are" in society, volunteering echoes the discourse of self-reliance and freedom from government involvement and assistance. When people without scientific credentials volunteer for citizen science projects, they may in fact facilitate the trend of shifting responsibility for environmental and health monitoring onto private individuals.

In the United States, environmental regulatory science has experienced significant budget stagnation and cuts over the years, which has encouraged the turn to citizen science. For instance, since the 1980s, when the EPA began to face severe budget cuts, the agency has promoted volunteer stream monitoring. According to the agency's website, the Office of Water "work[s] to expand the use of credible volunteer monitoring data at the federal, state, and local level."[19] The agency sponsors national conferences for volunteer organizers, publishes manuals on volunteer monitoring methods, and maintains a searchable database of volunteer monitoring organizations. Budgetary pressures also encouraged state regulatory agencies to rely on water quality data collected by volunteers. At least thirty-two states had citizen monitoring programs by 1992.[20]

The reduction of funding for public research and resultant reliance on volunteers is not limited to water governance. Consider the following case of participatory pesticide monitoring. In 2005, the California Department of Pesticide Regulation (DPR) started a twelve-month monitoring project. It selected the small rural town of Parlier, a part of San Joaquin Valley, with a 97 percent Hispanic population (median family income is less than $25,000/year). The local advisory group consisted of community members, environmental and farm advocacy groups, and others. These advisors were involved in the data collection in several ways, such as determining the types of pesticides to be monitored, the locations of air sampling monitors, and schools to be monitored. According to DPR, the project was "the first time a local advisory group played a key role in helping DPR frame goals, select monitoring sites, and decide other aspects of the project."[21]

A group of social scientists at the University of California, Davis analyzed this project, finding that even though regulatory scientists collaborated with citizens to monitor the problem, the proposed solutions to pesticide pollution did not directly confront the powerful pesticide industry. Regulators emphasized voluntary, market-based mechanisms, such as a voluntary code of conduct, rather than mandatory and government-based regulation.[22] The social scientists who studied this case wrote that citizen science was associated here not with "collective action and empowered citizenship" but rather with "techniques, especially biomonitoring, that tend to individualize environmental hazards and emphasize self-care as a key component of citizenship."[23] In this case, volunteer-based monitoring, albeit participatory and community based, did little to address the profound challenge of industrial pollution and the disproportionate effects it has on marginalized communities. Rather, it seemed to facilitate the neoliberalization of pesticide governance.

The rollback of government involvement has taken place not only in environmental protection but also in the realm of research funding. As government funding for scientific research has declined, government policies have encouraged citizen science as a means to generate low-cost data, and academic researchers have turned to volunteers to help get costly research projects done.

Budget cuts have prompted academic scientists to take a greater interest in data collected by volunteers. In 2017, scientists Marc Edwards and Siddhartha Roy pointed out that "the static or declining federal investment in research" has created the "worst research funding scenario in 50 years and further ratcheted competition for funding."[24] Reduced public funding for scientific research has often incentivized scientists to look for private funding (foundations, corporate donations, investors) and to seek ways to reduce the cost of research projects—including through volunteers and crowdsourcing. The use of volunteers as unpaid research assistants—in the form of citizen science—reflects this trend. There is an expectation that citizen science can help scientists survive the environment of funding scarcity to get science done.

Citizen science can offer volunteer opportunities for public-spirited individuals while enabling research that might otherwise be overlooked for funding. However, there is a strong argument to be made for stable public funding for professional environmental sciences. In a commentary on private funding of research, Marcia McNutt, the editor in chief of *Science*, wrote, "Without adequate federal support, gaps of all kinds can develop—in the balance of exploratory, basic, applied, and translational research; in the support of scientific talent at different levels of training; and in the support of different types of institutions."[25] Citizen science cannot fill all these gaps. Volunteers are difficult to recruit and retain. Their interests may be biased toward issues that have charismatic symbols (such as songbirds) and that reflect their immediate concerns (such as water pollution in their neighborhood), leaving some issues unaddressed. In basic scientific research on environmental issues, it takes a long time to accumulate baseline data, and it may be hard to sustain the interests and energy of citizen volunteers. In addition, given that there is a significant inflow of private corporate (and foundation) money into scientific research, the shrinking government funding may mean that scientific agendas are increasingly set by those private interests with significant financial muscle. It is hard to pursue an alternative research agenda if your only resources are small grants and volunteers.

Decreasing public expenditures for environmental science is a difficult challenge for both academic and regulatory scientists. Turning to volunteers to alleviate funding constraints makes sense in this context and may even lead to new discoveries and perspectives. However, embracing citizen science without also critiquing and resisting harmful changes in science funding policy suggests either giving up or becoming complicit with austerity policies. Scientists who work with volunteer researchers should be aware of this tension.

It is also necessary to ask who has the capacity to become a volunteer and who is most likely to be pressured to volunteer. While volunteerism is a value that seems universal, sociological studies on volunteering have shown that the availability of socioeconomic resources is an important parameter of volunteering. First, people's willingness and ability to volunteer is stratified by economic means. People may not get involved in volunteering because they cannot afford the expenses it requires. Another resource that is unequally distributed is discretionary time. This is not to say that longer work hours necessarily suppress volunteering; in fact, high-income people like lawyers tend to volunteer at higher rates. It is perhaps the flexibility of the time and location of paid work that changes people's ability to volunteer. People who have more control over their work hours and place are more likely to volunteer than people with a regular shift.[26] On the other hand, the expectation to volunteer might be stronger for some than others. Women may experience more social pressures to volunteer, particularly when the issue is framed as relating to children and education.[27] The use of women's free labor for nonremunerative work has been critiqued as exploitative and opportunistic.[28] Additionally, the social pressure to be "productive" and "give back" to society may target the unemployed, welfare recipients, and retirees.[29] These findings indicate the need to explore more thoroughly the nature of volunteerism in participatory environmental research. We will consider these tensions as they emerge in the case studies of the next three chapters.

In contrast to the vision of nonprofessional participants as volunteers, some environmental research projects conceptualize participants as activists and organizers for change. Important and influential forms of collaborative environmental research have arisen out of social movements for environmental justice. Environmental activists often face data gaps in regard to their environment and health status. In order to substantiate their claims about pollution and environmental illness, activists collect scientific data themselves, and this may be the only research that acknowledges the experiences of marginalized communities that are burdened with pollution. However, another common dilemma arises for activists who make scientific claims: they are often accused of "politicizing" science—tainting research with "bias" toward a political outcome. Scientists who work with activists may also become subject to similar criticisms.

The idea that science ought to be distanced from social influence has a long history in Western cultures and shapes environmental politics and policy today. Historian Robert Proctor documented the history of the notion of value-free science, tracing it from Plato to early modern philosophers to the present day.[30] In the United States, a "pure-science ideal" took hold after the Civil War, gaining momentum through the twentieth century.[31] In recent decades, both scholars and popular writers have debated the implications of "politicizing" science, suggesting the need to maintain distance between science and political processes even as they depend on each other.[32]

Today, experts frequently dismiss scientific claims made by activist groups because they assume activists' data are biased and inaccurate. Opponents often accuse activists of being "anti-science" if they dare to challenge the conventional way that scientific inquiry is done. And if scientists become allies of activist groups, they may be viewed as losing the neutrality that is presumed necessary to maintain their credibility. Within the scientific field, there are strong sanctions against activist-scientist alliances.

The prevailing idea holds that science is more credible when it is free from the taint of advocacy. Echoing this belief, an expert on research ethics, Nicolas Steneck, wrote in a background report for the American Association for the Advancement of Science that "when scientists become advocates, they become 'partisans' and are no longer neutral conveyors of scientific information. . . . While the line between neutral and partisan, between dispassionate and passionate, is not easily drawn, it nonetheless exists."[33]

The presumption that scientists should be "neutral" in their political commitments poses a problem for both activists and scientists—namely, how to maintain scientific credibility while taking a stand for justice. Gwen Ottinger, an expert on citizen science and environmental justice, summarizes the problem: "Because social movement-based citizen science is by definition political, it is often discounted or dismissed by scientists concerned that it is not sufficiently objective to make a reliable contribution to scientific knowledge; policymakers, similarly, may believe that it is not rigorous enough to be responsibly used to inform policy." However, she contends that the political leanings of a citizen science effort "should not disqualify it as a contribution to science or policy. On the contrary, given that value judgments are inevitable in all scientific investigations, the explicitly political nature of citizen science grounded in social movements suggests ways that all forms of citizen science—and science in general—could become more robust by being more transparent and more deliberate about their own values. Furthermore, by diversifying the values that inform scientific inquiry, social movement-based citizen science can help scientists identify fruitful new methods and avenues of investigation."[34] While a growing number of scientists may share this point of view, the idea that activism contaminates science and damages its credibility has exhibited strong staying power.

Citizen science projects that are rooted in social movements must tread cautiously on this complex political terrain. There is no straightforward way out of this dilemma; however, many people committed to important causes have navigated it before, and in the next three chapters, we will try to learn from their experiences.

When scientists work with citizens' groups, they have to balance the roles of an advocate and that of a scientist, which are frequently seen as incompatible with each other. Furthermore, from the perspective of citizens, using science to make their claims credible and reasonable might seem incompatible with disruptive politics. In a social context that pits science against activism and advocacy, citizen science has a fraught relationship with collective mobilization for environmental justice.

Putting Data in a Social Context

A virtue of citizen science is that it can help laypeople gain influence in environmental governance, but this presumes that decisions about our shared environment are based on expert knowledge rather than shared public concern. The high value placed on science-based decision-making is deeply ingrained in modern industrial societies that have experienced what sociologist Max Weber referred to as "rationalization." The process of rationalization replaces traditions, values, and emotions as motivators for behavior with concepts of reason, quantification, and scientific analysis. What would participatory research look like if it produced data that was richly contextualized with the history and culture of the people who collected it? And what would environmental governance look like if these forms of knowledge were treated as crucial to decision-making? Imagining such a possibility brings into focus another dilemma for participatory projects.

In many parts of the world, it is taken for granted that science should be the primary foundation for policies and government regulations. Sociologists have tracked the spread of "scientized" discourses and organizations around the globe, influencing politics, environmentalism, human rights, the economy, and other major institutions.[35] In many environmental governance regimes today, decisions are based on environmental risk assessment. This means that when social controversies arise concerning environmental pollution and health, policy makers and interested parties seek to resolve issues by relying solely on "neutral" facts and assessments.[36]

Science-centered environmental governance aims to produce rational, unbiased decisions—certainly an appealing goal in a time of heated political conflict over facts and truth. However, the rationalization of environmental issues warrants critical scrutiny. Most controversial environmental and technological issues involve scientific questions but are not reducible to technical matters; social, cultural, and ethical issues are also relevant. Rationalization also tends to privilege credentialed scientists as "experts" while marginalizing people who might bring other useful insights.

How does environmental citizen science fit into this context? When environmental decision-making focuses on technical information, citizen science can provide a way for ordinary people to participate. For example, participatory environmental research can educate citizens to become more scientifically literate, enabling them to better understand and make arguments about natural resources. Practically speaking, people without scientific expertise are at a considerable disadvantage in contexts where science is seen as the only legitimate basis for environmental and health governance; therefore, learning to do science can help people gain acceptance into decision-making processes.

Furthermore, rationalization can be an advantage in repressive political contexts. In undemocratic societies, focusing on scientific questions may provide a safe way for people to gain some influence in environmental decision-making. Consider the case of a citizen science project in a society that does not fully guarantee political and civic rights for its citizens. In Malaysia, a project called Kelab Alami trains young people—many of them high school dropouts or those with only an elementary school education—to identify marine species and report their observations. The fishing community and the marine ecosystem of concern are threatened by a massive real estate development project led by a Chinese developer in collaboration with a local investment firm. The planned "forest city" consists of four artificial islands and will have shopping malls, golf courses, hotels, offices, and high-rises for more than seventy thousand residents. People in the fishing villages have opposed the project, but the political and economic power behind it is too enormous to challenge. The sultan of Johor—Johor is one of the states where sultans, as constitutional monarchs, have considerable power—has invited the Chinese investor and reportedly has a strong financial stake in the forest city project.

In this context, opposing the development projects from the position of marginal fishing communities is politically not possible. Instead of taking a confrontational or oppositional stance to the project, the volunteer researchers work with the corporate social responsibility arm of the developers so the developers can show that they are being environmentally sustainable. They also work with the port authority because it also needs some baseline data on the marine environment. Through the medium of scientific data, Kelab Alami aims to raise awareness of the marine ecosystem, the fishing communities, and their valuable local knowledge so that the power holders can no longer ignore them. The success of this effort is still to be seen.[37]

The rationalization of environmental issues can generate yet another dilemma for citizen science. Focusing on collecting scientific data can diminish the transformative power of citizen science if leaders and participants lose sight of their broader social and ethical concerns. Participants may hope that their data will back up their environmental concerns, but data are never straightforward. When findings do not support the hypothesis conceived by participants, they may wonder whether there were problems with the ways that they measured and quantified the environmental problem of concern. Alternatively, when data do substantiate a community's concern, the data's pedigree and implications are always open to contestation by others.[38] Spending time on a "data treadmill"—striving to quantify issues that are "eternally requiring further verification"—may exhaust a community's energy and resources.[39]

Furthermore, turning to scientific analysis might sideline questions of environmental injustice. In a 1997 essay, David Pellow relates the experience of People for Community Recovery (PCR), a small grassroots environmental organization in Chicago. PCR formed in 1982 to fight for the environmental health of the ten thousand residents of a subsidized housing project located in what was known as the "Toxic Doughnut." Ninety-eight percent of the residents were African American, and their homes were surrounded by polluting industries and more than fifty landfills. PCR conducted a door-to-door health survey, discovering that residents were frequently ill with headaches, nausea, and other symptoms that they attributed to the fumes from the surrounding factories and dumps. The results of this first round of citizen-collected data were presented to the EPA. The agency responded by saying that they would need to see stronger evidence, so the activists convinced university public health researchers to carry out a more sophisticated health study. This second study convinced the EPA to begin an initiative to reduce pollution in the area.

Pellow observes that the case could be interpreted as an example of community empowerment through participatory research. The activists made progress toward their goal of reducing pollution. However, the case could also be interpreted as community dis-empowerment because, in rationalizing their assessment of the neighborhood's problems, the activist group was subordinated to scientific experts. The historical, political, and economic experience of environmental injustice—rooted in racism and poverty—was "reframed in the form of scientific documentation, mostly devoid of any sense of conflict or oppression." Pellow concludes that popular epidemiology (citizen health science) is "both costly and beneficial, empowering and disempowering."[40]

Rationalized descriptions and observations about the environment and human health can help articulate the experiences of pollution by regular people and give them a chance to sit at a table with credentialed experts, corporations, and regulators that are usually more adept at employing such technical and scientific language. At the same time, it could lead to a never-ending quest for more data points and sophistication and further validation and verification. As we will see in the case studies to come, by creating a data treadmill, rationalization can "foreclose the imaginative horizons of 'how' and 'why' in favor of 'how much.'"[41] Striving to contextualize data is therefore a key task for citizen science.

Shifting Spatial Scales and Time Scales

Environmental issues are often not contained in a single nation-state or locale. Even when pollution is localized, the chain of responsibility and causality might transcend national boundaries, requiring knowledge and actions at local, regional, and global scales. Citizen science can also take place over different timescales. Long-term monitoring, for example, works on a very different timescale than responding to a disaster. The question of *scale* thus poses a fourth dilemma for environmental citizen science.

While geographical scales may seem to be given and immutable, concepts like the "national scale" or the "local scale" are invented historically and only seem natural because of their frequent usage over long periods of time. The social shaping of scale is also evident, for instance, in suppliers' efforts to orient consumers toward the local or regional—as in "buy local" campaigns and the 100-Mile Diet— or in environmental groups' attempts to raise awareness of the watersheds in which people live, thus reorienting perceptions toward the ecological scale at which polluting activities have an effect. In each of these examples, people must imagine themselves to be citizens of nested or overlapping political communities—such as the nation, the state, the town, the watershed, and so on.

Citizen science can take place at multiple ecological and political scales, even spanning national borders. Consider what kind of participatory study might enable people to respond to natural gas emissions from faulty infrastructure. Emissions may be measured locally as air pollution or at a planetary scale using satellite imaging. The pollution itself may move across political boundaries, while the company responsible for the emissions may be headquartered in another country. Local or state agencies might not have the authority to rein in emissions; one might have to turn to national governments or even international treaties. This situation poses an array of dilemmas and choices for organizers of participatory research.

Advocates of neoliberal economic policies often argue for the devolution of state responsibilities to localized governments and individuals in order to weaken state power over industrial activity.[42] Citizen science sometimes facilitates efforts to shift environmental governance to the local scale by allowing individuals and communities to take responsibility for monitoring and managing environmental changes. In comparison to professional science, citizen science may be more attentive to local specificity and diversity; for instance, residents may be able to record subtle variations in air quality that official monitoring stations omit. Localizing environmental science can have advantages, producing data that is rich in details and in alignment with the specific needs of individuals and social groups. Yet research at this scale can also provide an illusion of environmental protection when more significant actions by governments are needed. The capacity of citizen science to attend to local experiences may falsely imply that the problems can be managed at a personal level, obfuscating the broader causes and dynamics of environmental and health issues.

Social movements often try to "scale up" their efforts by gaining national and international allies in order to increase their political power, find receptive venues, build coalitions, and put pressure on unresponsive governments. In citizen science, this may mean presenting data in ways that resonate with nonlocal allies and decision-makers. Citizen scientists seeking a solution to an environmental problem may need to address multinational corporations, intergovernmental agencies, or transnational scientific bodies. Often communities find that they must "speak the language" of professional science to be heard beyond the local scale, so citizen science may assist with building that capacity. Yet this may have the unwanted consequence of silencing the locally specific ways of understanding the environment and a community's needs. The next three chapters show how this tension arises in diverse contexts concerning a variety of environmental issues.

Timescales can also pose problems for participatory research. Citizen science may emerge as a disaster response, as we will see in the chapter about the Fukushima nuclear accident. The discourse of disaster and crisis may be helpful to citizen science projects in getting volunteers and donations as well as media coverage and public attention. The urgency and short time frame implied in portrayals of disasters and crises, however, may stifle attempts to address more long-standing structural issues.

Dilemmas of Environmental Citizen Science

This chapter has aimed to explain what we mean when we say that environmental citizen science is political. Citizen science is a complex idea. If we consider its component terms separately, each one has multiple contested meanings. What is science? Is indigenous or local knowledge science? Does science embody social values? The definition of *citizen* is similarly complicated. Should citizens be more like volunteers or activists? Are their rights bound by a nation-state or do they transcend it? Citizen science relates to resource allocation (funding for science and environmental and health protection), different ideas about good governance (small government and volunteering), and citizenship (how citizens make public claims and participate in community life). It also entails choices about the scales at which data are collected and policy-relevant actions are taken. These are all political matters.

We have identified four dilemmas that likely face participatory research projects. There is no doubt that there are many different kinds of citizen science that confront these issues in diverse ways. One can find a broad array of topics, types of volunteer involvement, organizational settings, and applications of the resulting data. Some projects are designed by academic scientists, whereas others are led by environmental movement organizations. Some projects are rooted in field-based sciences like public health and ecology, which have different disciplinary histories and frameworks than fields that emphasize more controlled laboratory-based studies and modeling. We do not think that the identified dilemmas are experienced in the same way across all these variations. Rather, we hope that focusing on issues and choices helps readers think about other ways of seeing diversity within citizen science.

The next three chapters will discuss examples of citizen science in the context of three important and controversial environmental issues: fracking for shale oil and gas, nuclear power, and agricultural biotechnology. In each case, we will focus on understanding the dilemmas that surround environmental citizen science. We observe, for reasons introduced previously, that volunteering is a double-edged sword. Volunteering helps get important research done but can position participants as "free labor" and may create patterns of unequal participation. We also show that activists using participatory research face a double bind: they need scientific expertise to help their cause, but science is typically perceived as tainted or biased if it is associated with a social movement. Participatory research provides a way for laypeople to engage in political disputes that involve technical information; however, the examples in chapters to come will further show that focusing on collecting scientific data tends to narrowly rationalize environmental problems, sidelining people's broader social, aesthetic, and ethical concerns. Finally, the cases will reveal dilemmas related to scale; while community science often occurs locally, the roots of environmental problems are often of distant or global origin. By openly acknowledging and grappling with these dilemmas, citizen science can become a more potent force for change.

3

Investigating the Impacts of Fracking

A man and a woman, both white and around retirement age, crouch by a slow-moving stream near a small bridge in Upstate New York. One uses small electronic devices that generate numbers for pH and conductivity while the other records these results on a clipboard. They remark on the consistency with last month's measurements and make a few more notes about the bank erosion they have seen on the opposite side of the stream. They will share these data with their team, the other people who are collecting data about their watershed. These records of stream health will be valuable if companies get permission to drill for natural gas in this area. The team members hope they will be successful in maintaining their state's ban on the methods used to extract gas from the shale deep beneath their feet (fig. 3.1).

Several shale formations across North America have been opened to oil and natural gas development since the start of the twenty-first century, and production has surged, surpassing other sources of these fossil fuels. In 2018, more than six million barrels of oil were pumped out of U.S. shale formations each day.[1] Sixty percent of natural gas produced in the United States now comes from shale.[2] Numerous companies are involved in exploring and producing these resources, including household names such as ExxonMobil, Shell, and Chevron as well as companies that may be less familiar, like Chesapeake Energy, EOG Resources, Diamondback Energy, Devon Energy, and many others.

FIG. 3.1. Volunteers collect water quality data. Photograph by Kirk Jalbert.

Developing these resources has been highly controversial not only because of the climate impacts of fossil fuels but also because extracting oil or natural gas from shale involves a combination of techniques that differs from conventional drilling. First, a hole is drilled down to the shale (sometimes a mile or more beneath the surface) and then across the shale formation horizontally. Subsequently, cracks are created in the shale by injecting a mixture of water, sand, and chemicals under very high pressure. This is known as fracking—a term that is now commonly used to refer to the shale oil and gas industry more generally. The sand props open the fractures, allowing oil or natural gas to flow out of the shale.

The environmental impacts of shale oil and gas development are significant and wide ranging. Millions of gallons of fresh water are used in the hydraulic fracturing process, often removing water from drinking water sources.[3] Along with the oil or gas, contaminated fluid containing large quantities of brine (salts), toxic metals, and radioactivity also flows back out of the well.[4] The liquid and solid wastes that are produced pose significant environmental and public health risks if not handled securely,[5] and there have been numerous incidents of accidental and intentional spills and other violations.[6] Furthermore, well water contamination resulting from drilling activities, potentially indicating the pollution of aquifers, has been widely observed and is now the subject of a variety of investigations.[7] Other concerns include air pollution, soil contamination, the increase in truck traffic that damages rural roads and pollutes the air with diesel fumes, and the risk of chemical spills at drilling sites and along trucking routes.[8] Impacts are not limited to the drilling sites. In Wisconsin, the source of much of the sand used in fracking, the landscape impacts and health hazards of sand mining have been extremely controversial.[9] There have also been numerous battles over the impact of building pipelines and other infrastructure on forests and wildlife, not to mention public safety. More broadly, fracking enables continued dependence on fossil fuels and delays investments in renewable energy and conservation.

Environmental citizen science has become an important force for documenting and resisting the many diverse harms caused by fracking, and this chapter examines a variety of these projects in the United States. Citizen scientists monitor different areas of concern, including water, air, and transportation routes. Some are primarily concerned with public health, while others focus on ecosystems, and still others aim to prevent catastrophic accidents. All these projects bring participatory research to bear on new and unanswered questions about the impacts of fracking and present a range of possibilities for changing the way that oil and gas extraction is governed in the United States.

Can environmental monitoring help people stand up to one of the most powerful industries in the world? To what degree do different citizen science projects address the inequitable distribution of benefits and harms? How can citizen science projects enable residents to have a say in the development of shale oil and gas resources? We pose these questions because fracking is not merely controversial; it is pivotal to American energy and climate policy today. The stakes are high for citizen scientists because their research contends with the implications of choosing to intensify oil and gas development rather than taking a renewable energy path. While many states and local communities have determined that fracking is an economic good, citizen science documents its environmental and health harms, bringing attention to the inequities produced by this industry.

In this chapter, we examine several examples of citizen science and look in detail at watershed monitoring, as it illustrates both the tremendous appeal of citizen science and the many dilemmas that participatory researchers are likely to encounter. Most of the primary research for this chapter took place in the Marcellus Shale natural gas play, where water-monitoring projects are widespread.[10] Watershed monitoring stands out for its remarkable geographic coverage and sustained commitment by volunteers. It illustrates the possibilities of productive partnerships among universities, national advocacy organizations, and grassroots community groups. Other citizen science efforts have made notable contributions as well. For instance, air sampling projects have brought attention to previously unknown public health risks and supported grassroots antifracking campaigns. Tracking of transportation routes has raised awareness of and mobilized opposition to the extensive network of trains that moves highly explosive materials across the continent.

Yet because of the public importance and urgency of research relating to fracking, citizen science projects encounter difficult dilemmas. For example, when volunteers mobilize to monitor their local streams, some may see this as letting government agencies off the hook for this crucial task. Another problem arises when deciding whether to take a stance against fracking. While some think citizen science and activism go hand in hand, others insist that science and politics should remain distinct. The examples explored in this chapter illustrate some of the ways that citizen scientists navigate these choices.

Volunteer Water Monitoring in the Marcellus Shale Region

The Marcellus Shale is a deep underground formation of marine sedimentary rock that extends throughout much of Pennsylvania and West Virginia, reaching also into large sections of New York and Ohio. In the early 2000s, high natural gas prices enticed a number of major oil and gas producers to explore the Marcellus, sending "land men" to convince landowners to sign contracts for access to shale beneath their properties. The governments of New York and Pennsylvania responded to the possibility of Marcellus Shale development in strikingly different ways. New York initiated a lengthy environmental review process and faced strong social opposition. Ultimately, New York state regulators did not grant the permits needed to develop the Marcellus Shale. In contrast, Pennsylvania quickly embraced shale gas development. The growth of Pennsylvania's natural gas industry was rapid. According to the U.S. Energy Information Administration, "Pennsylvania's natural gas production was more than eight times larger in 2015 than in 2010 because of development of the Marcellus Shale. Gross natural gas production exceeded 4.7 trillion cubic feet in 2015 and made the state the second largest natural gas producer in the nation, after Texas."[11]

As the industry grew, many concerned residents questioned industry assertions that fracking is "safe." From 2004 to 2016, Pennsylvania's Department of Environmental Protection (DEP) received nearly 10,000 complaints about environmental impacts of oil and gas development in the state. According to the reporters who obtained and analyzed the agency's records of these complaints, 4,108 were related to water quality problems.[12]

When drinking water (such as in a private well) is impacted, the DEP investigates and prepares a report for the affected resident, but surface water (rivers and streams) requires a different approach to monitoring. A variety of federal, regional, state, and municipal government agencies monitor surface water quality. Since the 1970s, state governments have been required to report to the Environmental Protection Agency (EPA) about the status of water quality under the Clean Water Act. In addition, the U.S. Geological Survey has sampled water quality nationwide for decades. Some government agencies also have begun new monitoring programs in response to shale gas development. For instance, in 2010, the Susquehanna River Basin Commission (SRBC), a multistate agency, began to deploy a network of fifty-one continuous monitoring stations that would specifically track surface water quality changes in areas of increased gas development activities. The Delaware River Basin Commission (DRBC) also created a monitoring initiative related to fracking.

All these efforts have been important, particularly for tracking long-term changes over time. However, in the context of the new threats posed by shale gas development, advocacy groups questioned whether these monitoring programs were aligned with public concerns. Pennsylvania's environmental regulators lacked the staff and resources to fully respond to public demands for fracking-related water testing. Realizing the shortage of resources, regulators themselves sometimes encouraged people to participate in volunteer water monitoring efforts to answer the questions that concerned them, as in this letter to a concerned citizen: "We currently do not have the resources to conduct baseline testing [of streams] prior to the start of drilling activities. The Department is responsible for assessing all of our waterways, and should therefore be able to document an impact that would actually cause impairment of a stream's designated use. However, it might be difficult to measure more subtle changes. We strongly encourage citizens who want to be involved in protecting their water resources to participate in volunteer monitoring programs."[13]

Seeking information about the impacts of fracking on their local watersheds, communities across the state of Pennsylvania—and in areas anticipating gas development in New York—launched volunteer water monitoring projects. By 2012, there were more than two dozen watershed groups monitoring impacts of fracking in Pennsylvania and several gathering predrilling water quality data in New York.[14] More than half of the watersheds in Pennsylvania's

Marcellus Shale region were monitored regularly by volunteers, while "baseline" monitoring was similarly widespread in the potential gas development regions of New York (fig. 3.2).[15]

Initially, there were no standard procedures for monitoring the effects of fracking, and because so little was known about what to expect, deciding on a monitoring protocol was a challenge. There are numerous methods of assessing stream quality, including visual observations, analysis of macroinvertebrates in a stream, and chemical testing. Some tools, such as conductivity meters, are electronic. Conductivity is a measure of a solution's ability to conduct electricity, which is directly linked to the total dissolved solids (TDS), such as sodium, calcium, and chloride. A spike in conductivity could be a sign that a spill of briny fracking wastewater has occurred. Other tools are chemical, such as titration kits

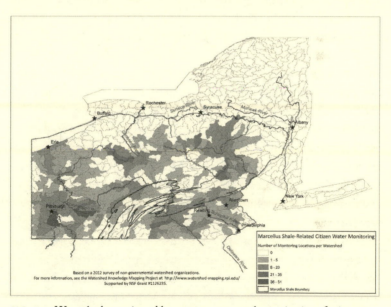

FIG. 3.2. Watersheds monitored by nongovernmental organizations for impacts of Marcellus Shale development, based on a study of watershed organizations working across the region in 2012. For an update to this map, see a report published by the FracTracker Alliance at https://www.fractracker.org/projects/water-monitor/spatial-and-social-inequalities/.

that measure dissolved oxygen. Some monitoring groups, such as the Pine Creek Headwaters Protection Group, located near the northern border of Pennsylvania, also trained volunteers to spot and report unusual industrial activities and changes to the landscape around the streams.[16]

Professional scientists working at both universities and nonprofit organizations became key allies to affected communities, in some cases initiating and coordinating environmental monitoring projects. For example, the Alliance for Aquatic Resource Monitoring (ALLARM), located at Dickinson College in Carlisle, Pennsylvania, developed a protocol for Marcellus Shale water quality monitoring. Previously, ALLARM had worked with watershed associations to gather data about acid rain. They were able to use this experience to develop a protocol and training in response to the new threat of fracking. In the fall of 2010, ALLARM released its "Marcellus Shale Volunteer Monitoring Manual." ALLARM's protocol uses TDS and conductivity as "red flag" parameters to indicate possible contamination, which then triggers the collection of samples for laboratory testing for "signature chemicals" whose presence can identify flow-back water from fracking (fig. 3.3). ALLARM chose barium, strontium, and total alpha as the signature chemicals to test in the lab. ALLARM went on to partner with a variety of organizations to conduct at least sixty workshops for volunteers across the state.[17] In New York, the Community Science Institute (CSI), based in Ithaca, did similar work, developing a monitoring protocol, training volunteers, and periodically testing water samples in a professional laboratory.[18] With the guidance of groups like ALLARM and CSI, volunteer water monitoring projects focusing on the impacts of fracking spread rapidly.

Abby Kinchy surveyed watershed organizations in 2012 to learn about their practices and goals. Most emphasized the virtue of education through water monitoring, aiming to inform the public about threats to the watershed. Most groups also wanted to contribute to scientific knowledge, and those with sufficient resources and training aimed to produce data that regulators and professional scientists would see as credible. Beyond informing the

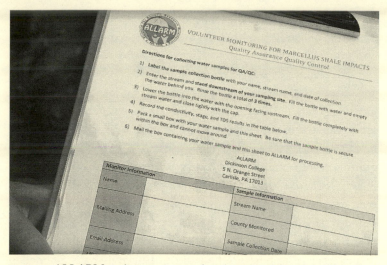

FIG. 3.3. ALLARM quality assurance and quality control data sheet. Photograph by Kirk Jalbert.

public and establishing credible scientific knowledge, most monitoring groups hoped that they would improve oversight of the natural gas industry by reporting observations of pollution or other damage to environmental regulators.[19]

Interviews with volunteer water monitors often revealed ambivalent attitudes about state regulators. It was widely acknowledged that government monitoring of water quality was inadequate. Yet at the same time, many trusted that if they had adequate water quality data, the state would act to protect public health and the environment. They saw their efforts as extending regulatory agencies' "eyes and ears on the ground" and believed that having volunteers in the field would increase the gas industry's efforts to avoid violations. The leader of one volunteer program explained, "We have had conversations with [a natural gas pipeline company] to let them know that we are watching their every step, that we have continuous monitors, we have volunteer monitors, we are going to do visual assessment along it. And that if you make a mistake, we are going to hold you accountable to it and make sure the [state] holds you accountable as well too."[20]

In 2016, the Dickinson College scientists who led ALLARM published an analysis of their Marcellus Shale monitoring project. The study reports that on at least forty-four occasions, volunteers visually observed and photographed pollution—mainly erosion and sedimentation—related to natural gas development. The article indicates that "the responsible agencies responded to the reports and took mitigation actions in a timely fashion." For example, "One pipeline observer captured mudslides in Tioga County, Pennsylvania, where the County Conservation District took the lead in addressing the situation with the company." Surprisingly, however, as of 2015, volunteers in the ALLARM network had not revealed any fracking-related pollution events by testing water chemistry. One reason is that most of the volunteers' measurements were in watersheds that did not have a shale gas well—thus they were considered "baseline" data. Another reason, the authors of the report observe, is that "volunteers are extremely cautious about reporting violations because of the contentious nature of the situation [and] actually may have under-reported probable incidents." Far from stirring up outrage about the impacts of fracking, volunteer water monitors may have held back from voicing concerns about unusual water testing results.[21]

Other Citizen Science Responses to Fracking

Volunteer water monitoring has been popular in the Marcellus Shale region, but it is not the only form of citizen science addressing the threats of fracking. In 2012, two national environmental justice advocacy groups, Global Community Monitor and Coming Clean, recruited grassroots activists to study air pollution around oil and gas development sites in six states. Residents of each state were trained to use "field log sheets ('pollution logs') that allow each resident to record what they see, hear, feel, smell, and taste in areas downwind of industrial activity as they go about their daily routines."[22] Then they took "symptom-driven" air samples at sites that they identified "based on the location of odors and health symptoms that have been experienced and reported on a consistent basis."[23] This was an important methodological contribution. Symptom-driven air sampling is an innovative way to trace connections between embodied heath experiences and environmental contamination and therefore has the potential to improve regulatory science.[24] The authors of a peer-reviewed paper resulting from the study explained,

> Symptom-driven samples can define the proper length of a sampling period, which is often limited to days or weeks. They can inform equipment placement for continuous monitoring and facilitate a transition from exploratory to more purposive sampling. Testing informed by human health impacts, and more precise knowledge of the mix and spacing of sources that may contribute to them, contrasts with state efforts, which are limited by access to property, sources of electrical power, fixed monitoring sites, and the cooperation of [landowners and oil and gas well] operators. In these ways, community-based monitoring can extend the reach of limited public resources.[25]

In other words, the knowledge that citizen scientists bring to a research project can be a crucial resource for improving regulatory science.

Once locations were chosen, trained community-based researchers took samples with an array of devices, such as air monitoring buckets. The buckets, pioneered by environmental justice activists with the Louisiana Bucket Brigade, are simple devices that capture air samples in special bags (fig. 3.4).[26] The study team also used SUMMA canisters (technical instruments used to collect air samples), formaldehyde badges (personal monitoring instruments), and dust wipes (tools to collect dust for analysis). Samples of air captured by the buckets and with the other devices were analyzed at a professional laboratory. The findings, while exploratory, were worthy of concern. In some samples, levels of volatile chemicals—including benzene, formaldehyde, and hydrogen sulfide—exceeded federal health guidelines. The results were first published in an online report and then in a peer-reviewed study

FIG. 3.4. Air-monitoring bucket built by the
Louisiana Bucket Brigade, ca. 2003. Photograph
by Gwen Ottinger.

in the journal *Environmental Health*, with community activists
among the coauthors.

There are now even more ways for community activists to
sample air quality, thanks to research efforts at a variety of universities, including international ones. The Speck is a device created at
Carnegie Mellon University and used by a nonprofit health advocacy organization, the Southwest Pennsylvania Environmental
Health Project (EHP), to provide users with real-time measures
of particulate matter in the air.[27] The Citizen Sense project led by
Jennifer Gabrys, a professor at Goldsmiths, University of London, provided a combination of sensors, including the Speck and
other devices, to residents of Northeastern Pennsylvania. Participants received digital tools for interpreting the data they collected
and created "data stories" about what they observed over time.[28]
In at least one case, measurements that a resident made with the

Speck prompted the EPA to carry out more extensive air quality testing.[29]

In another innovative project, Sara Wylie and her collaborators at the Social Science Environmental Health Research Institute at Northeastern University designed a low-cost method of detecting hydrogen sulfide using photographic paper.[30] Wylie began creating tools for citizen science as a graduate student at MIT, where she was part of a team that developed online resources to support communication about fracking. One of the websites, called Landman Report Card, enabled people to record their experiences when land men—gas industry employees who try to convince people to sign contracts with drilling companies—come to their homes. Based on familiar websites like Yelp or Angie's List, the idea was to "encourage community-to-community education so people could visit the site to find out more about the person at their door."[31] Another project was WellWatch, an online platform "where people could find out information about wells near them and geographically tag and share notes and complaints about those oil and gas wells."[32] Wylie and her collaborators went on to found the Public Laboratory for Open Technology and Science, or Public Lab, a nonprofit that encourages the creation of do-it-yourself (DIY) environmental monitoring techniques and devices. Wylie continues to design ways to record and interpret environmental health exposures and studies how communities use these new methods. In the case of hydrogen sulfide monitoring with photo paper, it appears that this tool is particularly beneficial for visualizing exposures to poisonous gases, complementing the sensory experiences of local residents.[33]

In some places, people are monitoring railroad traffic. Rail transport of oil extracted from shale formations in North America has increased dramatically in the last decade. As more trains full of highly flammable crude oil are rolling through population centers, the risk of catastrophic accidents increases—as tragically illustrated by fiery and sometimes deadly accidents in Quebec, Virginia, and other places along routes to oil refineries. Because train companies treat their shipping routes as trade secrets, leaving communities uninformed about the explosive cargo in their vicinity, public responses to the risk have included calling on citizens to monitor train yards. The nonprofit FracTracker Alliance and a research team at Carnegie Mellon University announced in October 2014 that they were seeking volunteers to record video footage of train cars passing near Pittsburgh, Pennsylvania.[34] In Washington State, volunteers with the Snohomish County Train Watch took shifts to count oil trains passing through their communities.[35]

Observing and mapping trains can be a powerful tool for advocacy to prevent environmental disaster. The Blast Zone Map, developed by ForestEthics (now called STAND), uses industry data, census data, and Google Maps to create an interactive website where users can find out whether they are in the "blast zone" in case an oil train explodes.[36] This tool is available as part of the organization's Stop Oil Trains! campaign, which seeks to reroute oil trains outside of population centers or halt the transport of oil by rail altogether.

Across the country, communities affected by fracking have participated in an array of citizen science projects. The multiplicity of citizen science efforts that can occur in a single community can be seen in this statement by Deborah Thomas, a community organizer with the Powder River Basin Resource Council in Wyoming (another state impacted by fracking):

> With Global Community Monitor's Bucket Brigade, we used tedlar bags and formaldehyde badges to sample air. Shale Test provided a FLIR gasfindir camera to see VOC emissions and summa canisters to collect air samples where emissions were seen. While interning with me, Cait Kennedy used real time monitors from Drew University to track particulate matter at compressor stations. Through collaboration with Sara Wylie at Northeastern University, we used photo paper film canisters to identify how the deadly neurotoxic, hydrogen sulfide gas moves across landscapes and through communities. Finally, we worked with the Coming Clean network, Commonweal, Shale

Test, and Subra Company, to collect air samples while we tested our bodies to see if the same chemicals would be found in both.[37]

Clearly, citizen science has been an important response to many of the environmental and health concerns surrounding shale oil and gas development. In some examples, citizen science projects support activism to prevent proposed wells or infrastructure or to document environment exposures. Other projects support regulatory efforts, focusing on preparedness for pollution events or reporting incidents to the authorities. What are the merits and drawbacks of these different approaches to citizen science? In the remaining sections of this chapter, we assess these projects by looking at how they contend with aspects of citizen science where dilemmas commonly occur. With each dilemma, organizers and participants in citizen science projects face a choice—yet the right way to proceed is rarely obvious.

Citizen Science Dilemmas

VOLUNTEERING

Addressing environmental problems through voluntary projects may inadvertently take pressure off of political leaders to develop necessary law and policy or to allocate public funds to crucial scientific research and regulatory monitoring. Accepting that environmental monitoring will be done by volunteers rather than tax-funded professionals often means that only some places—those where people are free to volunteer—deserve attention. In addition, projects that place volunteers into predetermined work roles (e.g., collecting monthly water samples) may fail to recognize the knowledge and innovative ideas that participants could bring to a project.

Citizen scientists face these dilemmas when they try to fill in gaps in regulatory science, as seen in the case of volunteer water monitoring in the Marcellus Shale region. Routine, long-term water monitoring is time-intensive and open-ended work, and it has historically—at least since the passage of the Clean Water Act in 1972—been the job of government agencies to do it. In the 1980s, when the EPA and other environmental agencies were facing severe funding constraints, water quality experts started to see volunteer watershed monitoring as a supplement to government monitoring efforts. At the time, conservation groups had already begun to track the problem of acid rain and other industrial pollutants with volunteer "stream watch" projects. By the end of the 1980s, the EPA was referring to voluntary research initiatives as a possible strategy to improve the nation's water monitoring capacity.[38] In a 1990 guidance document for states, the EPA identified these volunteer water monitoring programs as possible sources of knowledge that could satisfy public demands and could perhaps provide a low-cost source of data to cash-strapped agencies.[39] Interest in such programs rapidly grew, and by 2005 the EPA listed 872 volunteer watershed monitoring programs nationwide. In most states, there was at least one program that was approved by the state, the EPA, or both.[40]

In Pennsylvania, these ideas took shape in the Citizens' Volunteer Monitoring Program (CVMP), a state-run project initiated in 1996. CVMP provided training, equipment, and administrative support to eleven thousand volunteers in 138 watershed groups across the state.[41] It was disbanded in 2009 due to budget cuts, but many of the watershed groups continued their efforts after the end of the CVMP. When the potential impacts of fracking became evident, these groups had the knowledge, experience, and organizational skills to jump into action, seeking funding for fracking-related projects and training new volunteers to be water monitors.

In an interview with the leader of one volunteer water monitoring project, we raised the idea that government should be doing water monitoring, and the responsibility shouldn't have to fall to volunteers. The response likened the situation to the erosion of the social welfare state:

> Well, I would say it's the responsibility of the government to produce an economy that is capable of feeding everybody. But

instead, what we have is, like, 20% of our children in the United States living in poverty. And so how do they get fed? They get fed by people going out and donating food. This makes me mad, that it has to be that way. But that is the way it is. People go hungry unless other people step up. That is the government that we seem to have. The government does just enough to qualify as a government and then everything else has to sort of be done by the rest of us.[42]

In other words, the dilemma of volunteering is not limited to citizen science—It is widespread in a country where rollbacks to social welfare supports are frequently justified by saying it is the proper work of churches and volunteers to help those in need.

Even when volunteers "step up," however, gaps in services are likely. In the case of water monitoring, geographic coverage may be uneven because the capacity to volunteer is unequally distributed in society. Figure 3.2 shows not only the remarkable distribution of volunteering efforts across the Marcellus Shale region but also the remaining gaps. Many watersheds are not monitored by volunteers—perhaps because fewer people living in those areas have time, money, or organizational resources to engage in such projects. Addressing this problem may require statewide coordination, with resources targeted toward those watersheds, regions, and communities that lack sufficient capacity to monitor polluting industries. A study of one statewide program, Alabama Water Watch, found that the program mostly appealed to "more educated people with discretionary time and wealth," which left monitoring gaps in disadvantaged rural areas. The authors concluded that "significantly different approaches would probably be required to extend volunteer water-monitoring programs into resource-limited parts of the state." This might include partnering with organizations that focus on community development and livelihoods rather than the typical conservation or environmental advocacy groups, which are more prevalent in wealthier communities.[43]

What's more, volunteers face a political and cultural environment that devalues their work. They are told that participation in monitoring projects is empowering, yet they lack the authority to alter the polluting activities of the oil and gas industry. They are encouraged to develop scientific literacy, yet as we discuss later, scientists primarily value volunteers' knowledge contributions to the extent that the information participants collect can be compiled into large data sets. Conceptualizing citizen scientists as "volunteers" seems to place them into a role that is perpetually subordinated to "real" scientists. To their credit, organizers like ALLARM and Community Science Institute have pursued a deeply collaborative approach to working with volunteers. According to ALLARM, this means that the community sets the research agenda and collects the data, explaining that "volunteers work collaboratively with scientists on developing the study design and on data analysis, interpretation, and dissemination."[44]

Nevertheless, the notion of "volunteering"—which implies preconceived projects and work roles—may be an obstacle to developing models of environmental monitoring that can confront the complexity and unpredictability of developments like fracking. In Pennsylvania and New York, watershed groups could rapidly train volunteers in a variety of stream monitoring practices because there was already capacity for volunteer watershed monitoring in those states. However, many of the impacts of fracking—such as air pollution, groundwater contamination, and hazardous materials management, not to mention socioeconomic impacts—were not addressed through those citizen science projects. The volunteer stream monitoring model that was shaped around the problems of earlier decades (e.g., acid rain) channeled concerned citizens toward one kind of environmental monitoring without engaging the full range of concerns that arose from people's experiences and worries about fracking. There is a lesson to be drawn from this example: when one form of environmental monitoring becomes established over time, it is tempting to respond to new problems with those same tools and knowledge. Yet this may not be the most effective answer to the emerging issue. Tracking other impacts of fracking requires inventive approaches that do not fit easily into the existing volunteer framework.

Some other forms of environmental monitoring place less emphasis on recruiting and training volunteers, instead focusing on collaboration for problem solving. Air monitoring with "buckets," for example, is often targeted toward exploring a suspected cause of pollution linked to the illness experiences of local residents. Citizen science projects that build on the knowledge of affected communities, rather than the traditional model of volunteering to participate in a study designed by scientists, may be better suited to identifying the unexpected issues that arise with a new industry like fracking. For example, when new infrastructure is built, local residents may be unsure what questions to ask about its potential effects but could work together to compile their observations and decide what kind of environmental monitoring would be most appropriate to the situation.

TAKING A STAND

For many communities and organizations in Pennsylvania and New York, watershed monitoring was initially seen as a way to substantiate suspicions that the gas industry degraded local water sources. However, they faced a dilemma: any scientific claims they made might be perceived as biased if they appeared to have an antifracking agenda. How much time and emphasis should be placed on establishing the credibility of their scientific claims? Would taking a neutral stance bolster the credibility of their scientific work?

Gas industry representatives and supporters often publicly malign both professional scientists and volunteer researchers who speak out about the harmful effects of the industry. One pro-gas industry website attacked the "fake 'research'" being done by the EHP, described earlier.[45] Supporters of Marcellus Shale development often portray their critics (including academic scientists) as so biased against the gas industry that their claims lack credibility.[46]

Early water monitoring efforts encountered similar responses. A news article in the *Scranton Times-Tribune* recounted the reaction from Pennsylvania's natural gas industry, which characterized the Water Dogs as "vigilantes" who were likely to make false claims to regulators:

The Pennsylvania Oil and Gas Association posted news of the [Water Dogs training] event on its Web site under the headline, "Environmental vigilante training to help enforce Marcellus drilling regs." . . . Stephen Rhoads, president of the Pennsylvania Oil and Gas Association, said his post was meant to be "tongue-in-cheek," but the production of natural gas in the state and its regulation are "very serious issues." "To act like you're going to make someone competent as a DEP inspector with two hours of training for issues as complicated and arcane as these . . . is a little naive and a little dangerous," he said. He also said the volunteers risk creating more work for DEP, rather than helping its staff, "by reporting things that aren't in fact a problem."[47]

After this initial reaction, industry groups stopped criticizing volunteer watershed monitoring, perhaps deciding that these volunteer researchers were unlikely to pose a serious threat to industry operations.

It may be an industry talking point, but is "neutrality" really required for credible citizen science? The efforts to monitor air pollution stemming from fracking operations illustrate a successful strategy to maintain both scientific credibility and a commitment to environmental justice. Gwen Ottinger, an expert on citizen science, observed that the community researchers involved in the symptom-driven health study took care to establish their scientific legitimacy: "They partnered with credentialed scientists and they published their findings in a peer-reviewed journal as well as in an activist report. Furthermore, the article and report argue for the legitimacy of sampling results by drawing on the authority of the EPA: both refer to the EPA federal reference methods used to analyze samples, and the report describes how the sampling followed 'stringent quality control protocols originally designed with EPA Region 9.'"[48]

The oil and gas industry advocates attempted to cast doubt on the results by portraying the researchers as irredeemably biased—accusing them of politicizing science. Energy in Depth, a petroleum industry public relations website, published an article with the headline "New Air Quality Report Uses Scientifically Dubious Methods." Notably, the first several paragraphs of the article do not address the methods but instead focus on the anti-fracking positions taken by the lead organizations. The article contends that Global Community Monitor "could not be clearer that its focus is activism, not science."[49] The assumption here is that if the researchers have an interest in deterring fracking, their scientific claims cannot be trusted.

Despite the criticism—which was lacking in substance beyond calling attention to the activists' commitments—the impact of the study still stands. Results of citizen science were published in peer-reviewed journals, becoming a reference to others and potentially encouraging many more research studies.[50] This example suggests that there are steps that citizen scientists can take to bolster the credibility of their methods and to counter accusations that their political commitments diminish the quality and integrity of their research.

CONTEXTUALIZING DATA

In the previous chapter, we described the rationalization and scientization of many aspects of political life around the world. Decisions about new industrial developments, for example, are often made based on a scientific assessment of risks and benefits rather than traditions, aesthetics, emotions, equity, or ethics. People who turn to citizen science to address environmental problems in this context face yet another dilemma. On one hand, citizen science can provide a way for ordinary people to participate in regulatory processes—which are often closed to nonexperts. On the other hand, focusing on collecting scientific data can diminish the transformative power of citizen science if leaders and participants lose sight of their broader social and ethical concerns.

Shale gas development is a multifaceted issue, involving questions of energy infrastructure, economic exploitation, local governance, and the changing character of rural communities.[51] For many rural communities, the natural gas industry was a welcome development at a time of high unemployment and economic decline. Those who held the rights to the gas beneath their land were encouraged to sign contracts with companies such as Chesapeake Energy and Cabot Oil & Gas, often receiving lucrative signing bonuses, annual lease payments, and the promise of large royalties. Some believed the land men's assurances that they would barely notice the presence of a gas well after it was drilled. However, it was common for people to regard the industry with ambivalence.[52] Residents often observed that there were "winners and losers," or "some good and some bad." Indeed, industrial development troubled many residents, as summed up in this comment, shared in a focus group discussion: "My fear is that long term this industry is going to destroy most of the things that made this a nice place to live: low cost of living, low property taxes, very pure environment and clean water. We have increased traffic, crime, density of people. It will no longer be the rural, trusting environment that it has always been in the past."[53] Existing procedures for decision-making provided few opportunities to publicly address such broad concerns about the community way of life.

Ann Eisenberg, a legal scholar at West Virginia University, observed that landowners made decisions about allowing gas development without complete information. This led to both financial exploitation and exposure to environmental harms that were unknown to the landowners at the time of signing agreements. Eisenberg further observed that people and municipalities often signed agreements with gas producers under economic duress and that gas companies have engaged in bullying and predatory tactics.[54] Research by Stephanie Malin and Kathy DeMaster supports this claim: "Pennsylvania farmers we interviewed . . . described feeling constrained to make a 'deal with the devil.' They lost procedural equity. They complained of 'corporate bullying' in the face of large companies with 'rooms full of lawyers.' . . . As our study illuminates, we cannot assume that land ownership, or even mineral rights ownership, confers meaningful participation in farmers' decision making about natural gas operations on their land. Instead, Pennsylvania's small and midsized farmers experience procedural and other environmental inequities related to their participation in the natural gas industry and their growing dependence on those unstable leases. They feel limited by their economically vulnerable position."[55]

Citizen science could be a powerful challenge to these forms of environmental injustice, particularly if it creates opportunities for participants to advocate for themselves and their neighbors. But oftentimes, participatory research focuses on specific scientific measures of environmental pollution, which cannot fully represent the situations of social inequality. For many volunteer water monitoring groups, the goal is to gather scientific information that will be taken seriously by regulators as a means of informing decisions about the industry. Extensive efforts were made to develop good protocols for monitoring water quality, but parallel work was hardly ever done to help communities document their economic vulnerability and exploitation.

When I (Kinchy) spent a year as a volunteer with one water monitoring group in New York, I was struck by the participants' desires to scientize their concerns. Many volunteer water monitors were opposed to fracking. Some combined their environmental monitoring with more typical forms of activism such as protesting in the street. Yet in an early organizing meeting, some said that it was important to back up their "emotional" objections to gas development with "hard data," indicating that they believed the industry and regulators would take data more seriously than emotions. Others said monitoring the creeks would show the industry that they are serious about environmental protection. The volunteers sought to enhance their preparedness for legal battles over anticipated environmental problems by establishing baseline (predrilling) water quality. They planned to report findings to public officials and gas companies if pollution was identified and, if necessary, to collaborate with larger organizations to sue New York State or gas companies under the Clean Water Act. To support future legal action, they emphasized collecting scientifically credible evidence.

But would regulators really act on scientific data collected by volunteers? One example suggests otherwise. A controversy emerged in the southwestern corner of Pennsylvania, where discharge from an abandoned coal mine flows into Ten Mile Creek, a tributary of the Monongahela River. Volunteer researchers from the local chapter of the Izaak Walton League of America (IWLA), a conservation group, collected TDS readings that suggested the creek was polluted. The organization's leader, Ken Dufalla, also noted high bromide readings, a signature of fracking waste, in water quality data from the Water Resources Institute at West Virginia University.[56] Dufalla began to suspect that someone had illegally dumped fracking waste into the abandoned mine, which was draining into the creek.

In 2013, the IWLA volunteer researchers reported their observations and suspicions to state regulators, who followed up with their own tests in 2014, 2015, and 2016. The DEP's first analysis showed elevated levels of radium, a radioactive material that is present in the Marcellus Shale. However, further analysis by scientists at West Virginia University and later tests by the DEP indicated "acceptable" levels of radium, suggesting that that the DEP's first analysis may have produced a false positive.[57] Nevertheless, the volunteer researchers and other environmental advocates in the region remained concerned about other pollutant levels in the creek and questioned the DEP's sampling methods.[58] In 2015, Patrick Grenter, the director of the Center for Coalfield Justice, an environmental justice organization working in the region, said, "Recent testing has shown elevated levels of other shale gas signature pollutants [that] are harming public drinking water systems."[59] The DEP, however, maintained that the pollutant levels in this creek were not out of the ordinary, and no further regulatory action was taken.[60] This is an example of a data treadmill in citizen science, where environmental advocates strive to quantify issues that continually require verification—a task that can be demoralizing and draw focus away from broader issues at stake for the community.

Collecting scientific data in hopes of influencing regulatory decisions backfired in a different way in Erie, Colorado, where activists were fighting a proposal to drill for natural gas within

1,500 feet of two elementary schools. These concerned citizens participated in the multistate air quality study described earlier in this chapter. Using air sampling buckets to test air quality around other gas wells in the area, they found elevated hydrogen sulfide and silica dust levels.[61] These findings supported their efforts to question studies that suggested that fracking was safe. Yet rationalizing the opposition to fracking in Erie had some unexpected consequences. Local elected officials ignored their findings and treated the activists with disdain. Social scientists Skylar Zilliox and Jessica Smith observed, "Credible citizen science had the unintended effect of positioning activists as an uninvited or unruly public in the eyes of local government, threatening their ability to participate in the planning process."[62] Only after the community elected a new mayor and town board who welcomed a broader debate about fracking, one not merely focused on technical and quantifiable matters, did activists feel as though they were welcome to contribute to decision-making. This example suggests that even remarkably well-planned and peer-reviewed citizen science projects may not be received positively by local officials and that other forms of organizing—including electoral politics—may be more essential to creating local environmental protections.

Could an alternative approach to citizen science do more to help communities build power—without reducing their concerns to the narrow range of issues addressed by environmental regulations or getting on an exhausting data treadmill? The example of the Landman Report Card shows that environmental citizen science can be expanded beyond the natural sciences, enabling people to document and share sociological observations—in this case, the behaviors of gas industry representatives. While that project was experimental and short lived, it suggests that there may be more ways to develop participatory research that prioritizes the documentation and expression of social context.

SHIFTING SCALES

Community-based science can be effective at revealing localized impacts of polluting industries; however, the causes of pollution may not be local. If participants aim not only to understand but also to solve environmental problems, they may need to move beyond their local environments and make connections with activities that occur at national or even global scales. In the case of fracking, the immediate cause of water pollution may be an accident or careless work by a gas producer, but to actually hold the polluter accountable requires understanding the international scope of the industry and the relevant state and federal laws. In a more holistic sense, protecting watersheds from the impacts of shale development may necessitate a state- or nationwide struggle to reject fossil fuels in favor of more sustainable alternatives.

In the case of stream monitoring, volunteers typically focus on the local watershed, though larger networks of volunteers might regard entire river basins as their scale of interest. If problems are observed, attention turns to the scale of governance that is best suited to address it: Should town, county, state, or federal environmental officials be alerted? As seen in the example of the IWLA chapter that suspected illegal dumping of fracking waste in Ten Mile Creek, stream monitoring can get regulators to pay attention to problems that are happening at a hyperlocal scale—one point of pollution discharge among thousands of miles of streams in the state. The trade-off for this fine-grained localism is a loss of focus on the "big picture"—the overall impacts of shale gas development on the region's environment and public health. Volunteer stream monitors generally recognize that watershed impacts are just one piece of the larger set of problems produced by Marcellus Shale development, but making the connections in practice can be difficult.

Nevertheless, there have been efforts to aggregate watershed data in order to represent larger geographic scales. Some academic scientists have taken an interest in the data generated by volunteer watershed monitoring projects. Susan Brantley, a geochemist at Penn State, led an effort to facilitate data sharing about water quality in the context of Marcellus Shale gas development. The "Shale Network" uses free software called HydroClient that helps anyone search for, download, and visualize hydrologic data on maps that can scale from a single point to an entire continent. Through a series of annual conferences, volunteer researchers, staff of regulatory agencies, school groups, and academics have been trained to use HydroClient. Beyond data processing, the annual conferences enable participants to develop ideas collaboratively about how data sharing can improve understanding of the impacts of gas development. Another data aggregation project was developed at the Water Resources Institute at West Virginia University. That project, Three Rivers Quest, combines multiple sources of data, including volunteer researchers, on online interactive maps of three river basins that are impacted by gas development.

Does scaling up community-based research in this way help people pursue environmental justice? At this point, the possibilities remain speculative. Aggregating data across watersheds should allow researchers to analyze the impacts of fracking on wider geographic scales—for example, by asking whether water quality has declined in the state as a whole since the start of shale gas development. This could increase procedural justice—fairness and equity in decision-making regarding shale gas development—if it improves understanding of the potential consequences of decisions. Furthermore, it is possible that the data could reveal a pattern of environmental degradation that disproportionately affects certain populations, such as the rural poor. However, Shale Network leaders reported in 2018 that "the database did not reveal much evidence of contamination" even though regulatory reports about spills of hazardous fluids suggested that water contamination was likely.[63] The network thus continues to work to improve its methods of data collection, exploring many new sources of information beyond citizen science. Therefore, the outcomes for environmental justice are elusive.

The differing scales at which gas producers and volunteer researchers operate represent another intractable problem. While gas producers like BP and Chesapeake Energy make decisions based on their understanding of global energy markets, volunteer water monitoring groups generally view fracking at a more localized scale, such as concern for a fishing stream, a farming community, or local economic development. The industry has resisted local governance precisely because such "local" interests can be an obstacle to their global business plans. However, stringent state and federal regulations would be far more effective to limit the overall harms from the industry. New York State—and some countries, such as France—have banned high-volume hydraulic fracturing. If shale gas development is a threat to the environment and public health, then action at a state or national scale may be preferable to local interventions.

Volunteer environmental monitoring can support state and national campaigns by focusing on developing political advocacy skills as well as scientific skills. Citizen scientists are not only observers and stewards of their local environments; they constitute a political force for policy change at multiple levels of governance.

Conclusion

The expansion of the oil and gas industry into shale formations across the United States has been met with a parallel growth of grassroots and increasingly coordinated opposition. Citizen science is part of this story of resistance to a powerful industry. In an array of diverse examples, concerned citizens responded to gaps in knowledge about the impacts of fracking and sought to document the changes happening in their environments and communities. Professional scientists and advocacy organizations facilitated these projects and in many instances helped participants compile their findings in ways that transcended the local scale.

Because the threats posed by fracking are novel, citizen scientists often worked with innovative research methods. Some iterated new techniques for long-standing models of volunteer watershed monitoring. Others used novel devices to monitor air quality. Still others attempted to record unusual kinds of data, such as transportation routes and interactions with land men.

The outcomes of these projects are wide ranging. Some mobilized volunteers to be "eyes and ears on the ground" when government monitoring was insufficient, allowing for local regulatory responses to pollution incidents. Some led to discoveries that are written up in peer-reviewed publications. For others, the outcomes are less tangible but still important, such as stronger ties among participants or a proof of concept for an entirely novel approach to participatory research.

But our aim here is not just to tout the merits of participatory research; we probed these examples in order to think more carefully about the social context and politics of citizen science today, which often manifest in the dilemmas that we identified earlier. We saw, for example, that when volunteers are deployed to monitor the environmental impacts of corporate activities, notable gaps in coverage can be left because of the uneven availability of willing volunteers. Citizen scientists may also find themselves maligned for appearing "biased" against the industry and must make careful choices about how to establish their credibility while advocating for change. There are also problems associated with presenting concerns about fracking in scientific terms. We saw that measuring pollution can lead to a frustrating data treadmill and can intensify (rather than transcend) political divisions. Additional issues emerge when deciding on the scale of a participatory research project. On one hand, participants' knowledge and commitments may be greatest at the local scale, yet a local study may be inconsequential to an industry that operates at national and global scales. On the other hand, gathering data using standardized methods in order to aggregate it at a regional or state level may diminish attention to the particular experiences and environmental changes in different localities.

Citizen science in the context of the U.S. fracking industry may be unique; few other industries have had such rapid growth with so little federal oversight and in such close proximity to residential areas. Fracking has been banned in some countries, and many other governments are seeking to reduce the use of fossil fuels.

Other industries will have different dynamics. Nevertheless, this case highlights challenges that are likely to be relevant to volunteer researchers facing any new and previously undocumented source of pollution. Context matters—we can't make sweeping generalizations about what participatory environmental research can and can't accomplish. But with this case, we have begun to grapple with the dilemmas that arise and the choices that practitioners of citizen science must make.

4

Detecting Radiation

Ms. Tsunoda volunteers at a citizen radiation measuring organization (CRMO, commonly known in Japan as *shimin hōshanō sokuteijo*). It was set up after the Fukushima nuclear accident to provide food testing services to citizens who were concerned about radiation contamination. Today, Ms. Wada from the neighborhood has brought in a bag of spinach that she bought from a local supermarket. Ms. Tsunoda takes the spinach and records Ms. Wada's name and contact information. She makes sure that there is a sufficient volume of spinach to be measured. She then chops the spinach finely, puts it in a plastic beaker, and places it inside a scintillation detector. After several hours, the machine produces an estimate of the concentration of radioactive cesium. Ms. Wada is relieved to learn that the estimate is "N.D." (not detectable). Although she has been a volunteer at the CRMO since 2012, Ms. Tsunoda says she is not an expert in radiation measurement. In fact, she describes herself as a regular housewife; she only went to junior college and majored in English. She learned how to operate the detector only after the nuclear accident.[1]

The Fukushima nuclear accident occurred on March 11, 2011, after a massive earthquake and tsunami hit the northeastern part of Japan. The earthquake and tsunami caused several nuclear reactors in the Fukushima Dai-ichi Nuclear Power Stations operated by Tokyo Electric Power Company (TEPCO) to lose electricity, disabling the cooling systems. The country was still barely coming out of the shock of the earthquake and tsunami when it started to recognize the problems at the nuclear reactors. Three reactors partially melted down.

Concerns about contamination and radiation exposure quickly emerged. The monitoring posts by TEPCO and the government showed higher-than-normal levels of radiation in multiple locations. Food contamination also became a concern. On March 19, the government announced that it had found contaminated food and subsequently ordered the governors of four prefectures to suspend shipments of spinach and milk. The contamination was not limited to Fukushima prefecture. The Tokyo prefectural government found that one of its water sources was contaminated by radioactive iodine and started to distribute bottled water for infants in the areas that received water from this source. People became increasingly concerned as the media continued to report "more and more food found above standards" and "25 out of 45 Fukushima vegetables above radioactive standards."[2]

In the first weeks of the accident, there was very little information that was helpful for regular people to understand the extent of contamination and what they could do to protect themselves. Even after the government started to test the air and food items, many felt the need for credible information. Residents were unsure about whether their neighborhoods were contaminated and to what degree. Monitoring stations were sporadic and concentrated around nuclear power plants. In terms of food safety, the government insisted that food was tested properly and that foods on the market were fit for human consumption, but many citizens felt that the government testing covered too few items.

It was in this context that regular people started citizen science projects, such as Ms. Tsunoda's CRMO, to monitor the contamination. There have been several major citizen science efforts after the Fukushima nuclear accident. CRMOs focused on food and sometimes beverages and breast milk. Another project, Safecast, focused on air quality. In North America, volunteers sampled ocean water to determine the level of contamination arriving from Japan. These projects are significant not only because they address urgent public health questions but also because they provide customized information that addresses residents' concerns about particular foods and locales.

Beyond providing customized information, can citizen science bring more accountability and transparency to the nuclear energy industry? Participatory studies may shed light on the uneven distribution of the benefits and harms of nuclear power production. So what are the ways that citizen science enables marginalized groups to gain a voice in nuclear energy governance? These questions are important to consider because the Fukushima disaster was not just a momentary crisis. The lack of data and information in the aftermath of the earthquake and tsunami in March 2011 was the most immediate concern, but many citizen scientists also viewed the crisis in a broader social and political context. The disaster crystalized many of the issues that critics of nuclear energy had been pointing out for decades regarding secrecy and lack of accountability.

The government and the industry emphasized the extraordinariness of the 2011 earthquake and tsunami to insist that other reactors would be safe. However, many smaller accidents had previously taken place, including an explosion in the Tōkaimura reprocessing facility in 1999, which killed several workers. Postaccident investigations by different commissions pointed fingers to different causes of the Fukushima nuclear accident, but the case is reflective of the general vulnerability of Japanese nuclear power plants that are old, and built in earthquake-prone areas.[3]

The underlying concerns with nuclear energy also go beyond the risks of accidents to include unequal distribution of benefits and harms. Even when reactors are in regular operation, nuclear power produces contamination and injuries. These are often hidden from public view, as Gabrielle Hecht deftly demonstrates in studies of French and African nuclear history.[4] From the extraction of uranium, to regular cleanup of reactor facilities, to the disposal of spent fuels, nuclear energy operation is inseparable from hazardous exposures, but these costs are usually invisible, since they are disproportionately borne by the socially marginalized.[5]

In this chapter, we draw on extensive field research in Japan to look at the array of citizen science initiatives that emerged in the aftermath of the Fukushima disaster. Using citizen science, concerned residents of Japan have been able to detect and reduce their exposures to radioactive material, often questioning official assurances of safety. As in the fracking case, citizen scientists who study the impacts of the Fukushima disaster encounter dilemmas. Questions of scale are particularly acute; projects range from individualized radiation monitoring (personal exposures) to monitoring at the community or regional scales. Without citizen science that provides fine-grained data, Japanese residents may have no other way to avoid radioactivity in their foods and environments, since the pattern of radiation contamination is highly uneven. At the same time, the individualized nature of radiation monitoring risks being complicit with the efforts to reframe nuclear safety as an individual responsibility rather than an obligation of the government and the nuclear-related industries. This is just one of the dilemmas that citizen scientists have had to face; others involve the roles of volunteers, the boundary between science and activism, and the risk of reducing nuclear politics to debates on a set of radiation standards. Awareness of these issues makes it clear that radiation-related citizen science is never simply about measuring radiation but is always embedded in the broader arena of nuclear politics.

Fukushima Citizen Scientists: The Case of CRMOs

CRMOs work as measuring "stations" where citizens can bring (and/or send in by mail) their own foods of concern to get the estimated concentration of radioactive cesium and iodine. The expansion of these organizations was impressive after the Fukushima nuclear accident, but they did not suddenly emerge following that crisis. After the 1986 Chernobyl nuclear accident, concerns about contaminated imported foods from Europe prompted the establishment of CRMOs in Japan. One CRMO in Koganei city in Tokyo started that way and continues to operate to this day.

In the early weeks of the Fukushima accident, there was little information about food contamination. But as time passed, the government expanded the screening system and started to post the testing results online. Today, foods from areas in and around Fukushima are regularly screened by the producer organizations and the government, and the data are made public. But even into the second, third, and fourth year after the accident, people established CRMOs to measure their own foods.

One of the main reasons for starting the CRMOs was strong distrust of the governmental agencies, the industry, and even scientific experts in the fields that were related to nuclear energy. Japan started to pursue nuclear energy in the 1950s, and as the construction of commercial reactors advanced, social opposition grew in host communities in rural areas as well as among urban critics. Activists and journalists exposed many instances of financial and political pressures that were mounted upon the residents of areas targeted for reactor construction. Investigations also revealed the utility industry's data concealment and falsification and the insufficiency of government oversight.[6]

Critics used the term *nuclear village* to describe the collusion of the business sector, regulators, and scientists, denoting the close relationships among the elites whose interests aligned in nuclear energy promotion. These elites included many professors in the nuclear physics departments of top universities in Japan, who had been working closely with the government and the nuclear industry. For instance, the Atomic Energy Commission, the organization that shaped the overall direction of nuclear policy in Japan, had been promoting nuclear power as a clean and reliable energy source. Its chairs were usually filled by engineering professors of top universities, playing the role of "cheer leaders for nuclear industry and nuclear power businesses," according to one observer.[7]

Experts who were not part of the nuclear village and dared to critique nuclear energy came to constitute an important part of the antinuclear movement. For instance, Takagi Jinzaburo (1938–2000) used to teach at a university but quit his job and helped establish the Citizens' Nuclear Information Center in 1975. The center is still active and plays an important role in the Japanese antinuclear movement. Nonmainstream experts like him collaborated with other antinuclear activists who critiqued scientific data and expertise emanating from the nuclear village. They highlighted the nuclear industry's history of data concealment and falsification, which served to downplay risks, and how mainstream scientists were implicated in these activities.

These critiques struck a chord with many people after the Fukushima nuclear accident. The government was slow to order evacuations, and even when it did, the scope was limited to residents in a twenty- to thirty-kilometer radius from the reactors, while the U.S. government ordered the evacuation of American citizens living in an eighty-kilometer radius. The Nuclear Safety Commission did not release the data from the System for Prediction of Environmental Emergency Dose Information (SPEEDI), which simulated the flow of radioactive materials, for more than a week. The government, the industry, and some of the mainstream scientists denied that the reactors had gone into meltdown and only later admitted it after mounting pressure to tell the truth.

In regards to food safety, the government insisted that the food on the market was safe to eat. It started the Eat to Support campaign, promoting the foods from the affected areas and commending their consumption as charitable and considerate of the disaster victims. Simultaneously, the mainstream media and the government policed food safety concerns, calling them *fūhyōhigai* (harmful rumors) that were baseless and harmful to the disaster victims and affected areas. Fūhyōhigai refers to declines in sales of products from certain areas because of a false belief that the products might be contaminated. The nuclear village used the idea to portray any expression of consumer concern with food contamination as tantamount to fūhyōhigai and thus the cause of suffering and loss of economic vitality for the people in the affected areas.[8]

Japanese women tended to be the target of these accusations because they do most of the food shopping and preparation in the home. The criticism was facilitated by the long-standing stereotyping of women as emotional and weak on scientific issues. For instance, women who were concerned with radiation were critiqued on social media as *nō-mama* (radiation brain moms). The term was a pun on *hōshanō* (radiation) and *nō* (brain), connoting mothers who were obsessed with radiation. Reflecting a sexist stereotype of maternal overreaction, mothers who raised concerns about radiation contamination were chastised as having a different kind of brain—one that was unscientific and unthinking. As anthropologist David Slater and his collaborators observed, postaccident Japan saw a widespread "medicalizing discourse that cast radiation fears as the result of individual pathology" and that particularly targeted women.[9] Sexist policing in the name of preventing fūhyōhigai was widespread after the Fukushima accident.[10]

It was in this context that citizen science projects started around the country to help regular people—often women—to measure their own foods and the environment. People wanted to know that their environments and the foods they were eating were safe, and they wanted to do the measurements by themselves because they distrusted the established experts and the government. According to the government data on radiation food contamination, for instance, about 2 percent of rice, 3 percent of vegetables, and 20 percent of mushrooms and fisheries products exceeded the government standard of 100 Bq/kg in the fiscal year 2011 after the accident, and the rate has declined since then. In the fiscal year 2013, less than 1 percent of vegetables and rice and 2 to 3 percent of mushrooms and fisheries products were above 100 Bq/kg.[11] From the perspective of the government, the standard of 100 Bq/kg is strict, and worrying about a small amount of radioactive materials below this level is nonsensical. Yet many continued to distrust the official account. CRMOs continued to expand, and as of 2013, there were seventy-four active organizations.

How do CRMOs measure different foods? Typically, these organizations use what is called a scintillation detector (fig. 4.1). Detectors are costly; the cheapest one can be obtained for around $10,000. Some nonprofits donated the detectors to citizens' groups, and other CRMOs obtained funds by collecting donations.

People who established CRMOs were quite diverse, including parents, religious organizations, technology hobbyists, activists on different issues (peace, environment, antinuclear, disability, social welfare, etc.), and "regular housewives." The vast majority of people at these organizations were not trained in nuclear physics or radiation measurement, and it was only after the accident they started to learn how to operate the detector and interpret the results. Some learned about radiation measurement by going to workshops, asking questions of other CRMO members, learning from the manufacturers and importers of the detectors, and reading blogs and

FIG. 4.1. CRMO staff member and detector (cylinder behind water bottles). Photograph by Aya H. Kimura.

websites. There were a few academics who gave technical advice to CRMOs as well.

CRMOs provided valuable data on food contamination (fig. 4.2). They tested more than sixty thousand food samples, including vegetables and fruits, meat, milk, and processed foods. CRMOs tended to use a lower minimum detectable level than the government's monitoring system. Epidemiologically speaking, the level of contamination that these organizations flagged might not increase cancer death in a statistically significant manner. However, the analysis helped people who wanted to avoid even low levels of radiation—for instance, parents seeking to minimize their children's exposure to contamination.

CRMOs also played an important role in educating citizens about measurement processes. For instance, they generally provided more information and explanation to clients who brought in food to be tested than the government-run measuring programs did. At a measuring program operated by the city of Fukushima,

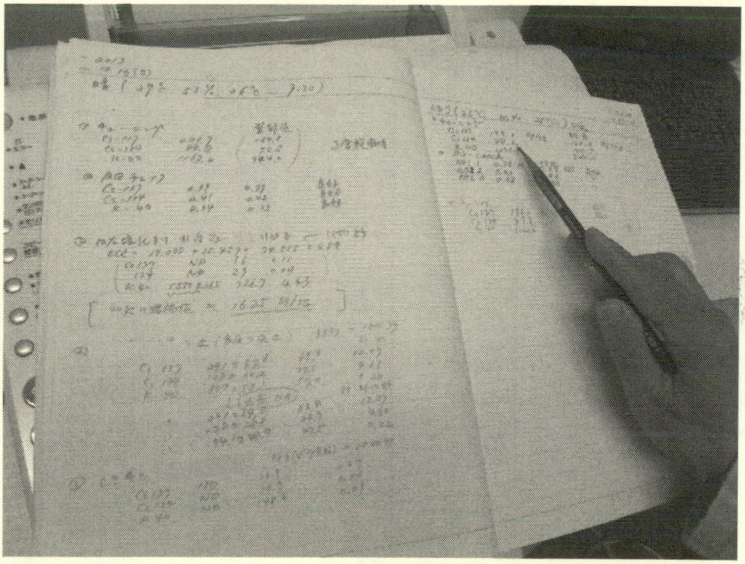

FIG. 4.2. Handwritten measurement log of a CRMO. Photograph by Aya H. Kimura.

for instance, citizens would simply drop off an item to be measured and come back in thirty minutes or so to receive a printout. A sample printout would read, "Dear Mr. X. In regard to your request for a measurement of radioactive materials on X/X/2013, it is as follows," with a simple table of three rows indicating densities of cesium-134, cesium-137, and the total of the two. It would say nothing about the process of measurement except for the name of the detector ("Belarus ATOMTEX Co. NaI Scintillator"), and it would have no graph showing a spectrum that is critical to interpreting estimates of radioactive materials. In contrast, most CRMOs provided detailed information about the testing process. Many had a folder in the office that was used to explain the detector and measurement process. Part of the reason was that staff members wanted to demonstrate the validity of their results by showing the details of the measurement, but they also wanted to educate other citizens about the science of radiation measurement.

These citizen organizations also provided a social space for people to share concerns. For instance, a CRMO run by a consortium of Christian churches recognized that citizens need not only the data results from measurement of food but also someone to talk to about their concerns about food contamination and health risks. They decided to hire a staff member experienced in counseling. This CRMO prepared a comfortable space with a cozy sofa and plenty of room for people to sit and talk. While having a counseling professional on staff was unusual among these organizations, many of them envisioned their role as not only measuring the level of food contamination but also making a space for citizens to share their worries and concerns. This function was important, as one woman put it, because "people cannot do it outside." Many of the CRMO staff members interviewed talked about how people came and confessed their relief at being able to admit they were worried about radiation contamination, since the government and scientists emphasized the safety of food and criticized people who questioned it. One mother said in an interview, "Even though I was worried, I could not really say that in front of others," but when she came to a CRMO, "I could talk about radiation."

CRMOs functioned as a space for sharing concerns without having to fear being labeled "crazy" and "overly nervous."[12]

Other Citizen Science Responses to the Fukushima Accident

CRMOs primarily focus on food, but there are other types of citizen science projects as well. Safecast, for instance, is a loosely networked group of people that was established after the nuclear accident to crowd source data on airborne radiation.[13] The core people are versed in information technology and software design. They include academic researchers Pieter Franken and Joi Ito, who are affiliated with MIT Media Lab; Azby Brown at Kanazawa Institute of Technology; and people at the Tokyo Hackerspace. Recognizing that there was a lack of information about the impacts of the nuclear disaster that was unfolding in front of their eyes, they first tried to map radiation by using existing Geiger counters. Realizing the need for scaling up, they then devised easy-to-make Geiger counter kits and a web-based mapping platform where laypeople could upload their measurement data, which were time stamped and tagged by location. Similar to many CRMOs, Safecast operates on the principles of participation and openness. The software and hardware are open source, created through the collaboration of many people. Volunteers can purchase Safecast's Geiger counter kits and upload data to a shared website.

Safecast's website now shows data measured at twenty-seven million locations mapped geographically, where the user can zoom in and out to check a particular neighborhood.[14] By 2015, more than five hundred volunteers with Safecast's latest Geiger counter, bGeigie Nano, had sent in more than one million radiation data points. Safecast now includes data from at least seventy countries around the world.[15]

According to the organizers, the most important function of the group's activities is

to enable people to easily monitor their own homes and environments themselves, and to free themselves in this way from dependence on government. The impact this has had on people is overwhelmingly positive, though the responses that follow depend on individual temperament and circumstances. For some individuals, being able to take reliable radiation readings themselves, and to find radiation levels lower than feared, for instance, brings a sense of relief. For others, their findings serve as reliable confirmation of injustices suffered and risks imposed unnecessarily. Still others consider the results to be objective and useable confirmation of the degree of radioactive contamination they must deal with, supplanting both excessive fear and excessive hope, and enabling them to take informed action appropriate for their situation.[16]

Safecast therefore strives to provide objective, credible data that are customizable to varying needs.

In another project, several groups—some of them starting with a focus on food—are now mapping soil contamination data. A consortium of twenty-eight CRMOs started to map soil contamination in eastern and northeastern Japan in 2016. They were motivated by the fact that the government focuses on airborne radiation and largely ignores soil contamination, except in narrow geographic areas. In contrast, these CRMOs pointed out, when the Chernobyl meltdown affected Belarus and Ukraine, those governments used soil contamination data as one of the criteria for issuing evacuation orders. Soil contamination data might strengthen the argument that the Japanese government policy on evacuation and decontamination was too lax. In addition, without soil contamination data, the full extent of the consequences of the nuclear accident will forever be unknown. These groups wanted to document the geographic reach of the disaster before traces of the contamination disappeared, so they decided to launch a soil contamination mapping project. They crowd sourced funding and collected soil samples from seventeen prefectures.[17]

The Fukushima nuclear accident prompted citizen science projects outside Japan as well. A project called InFORM is headed by scientists affiliated with universities in Canada. Starting in 2014,

InFORM organized volunteers to sample ocean water at sixteen locations along the coast of British Columbia. Volunteers collected twenty-liter seawater samples and recorded important properties like temperature and salinity. The samples were then sent to university labs for radioisotope analysis. In the United States, similar ocean monitoring is organized by chemical oceanographer Ken Buesseler at the Woods Hole Oceanographic Institution. Under the project name Our Radioactive Ocean, Buesseler organized volunteers to collect ocean water on the West Coast. He also crowd funded the project. People can propose a location for monitoring and raise $550 to $600 for the cost of a test kit, shipping, and laboratory analysis. Both InFORM and Our Radioactive Ocean worked with volunteers and shared data publicly, but unlike CRMOs and the soil testing groups, the scientists played a central and defining role in the design and management of the projects.[18] They summarized the collected data in the academic journal *Environmental Science and Technology* in 2017. The article concluded that cesium levels did not threaten human health or the environment.[19]

CRMOs, Safecast, and soil and ocean monitoring efforts represented diverse citizen science activities even as they responded to the same event. All of them provided useful information and data. The ocean monitoring project provided radiation information for people in the United States and Canada who were concerned with Fukushima-related seafood/seawater contamination when no government agencies or international organizations were monitoring the accident's impacts on ocean water in North America. CRMOs, the soil monitoring project, and Safecast were able to respond to Japanese residents' needs for fine-grained data. The history of distrust of the nuclear village made these "citizen" data useful for many to understand the status quo of contamination despite the fact that official sources released more information as time passed. Citizen-produced data were important alternatives or additions to official information sources, and these organizations functioned as watchdogs and reference points.

After the Fukushima disaster, citizen scientists around the globe organized to address several areas of concern. Although the diversity of radiation citizen science makes it difficult to generalize, we hope to tease out dilemmas and tensions by thinking about its broader social and historical context.

Citizen Science Dilemmas

VOLUNTEERING

There are many benefits to voluntary radiation measurement. It is helpful to regular people, as they can get locally specific information when contamination is spread unevenly—an issue that is particularly notable with radiation. Additionally, when the official sources of data are deemed untrustworthy, regular people may view voluntary monitoring as a source they can trust. In this way, voluntary radiation measurement seems to bolster individuals' rights to know in an uncertain environment. However, to what extent can self-monitoring replace public environmental monitoring when we consider volunteers' access to resources and consolidated data? Not everyone has the capacity to volunteer, and individuals might not have access to sufficient information to make sense of the broader pattern of contamination and exposure.[20] Without connecting individual situations to the systemic drivers and dynamics, voluntary radiation monitoring risks obfuscating the need for more universal monitoring and radiation protection.

As we have pointed out in previous chapters, volunteering is often equated with empowerment, as it gives the feeling of self-efficacy. Voluntary radiation monitoring of food and the environment can be considered empowering, as it gives regular people a fuller understanding of their surroundings, enabling them to take action to protect themselves. Such sentiments were recorded in many interviews with citizen scientists. Many CRMO volunteers said that they could no longer trust the government or experts and therefore found self-help appealing. To them, the government seemed more intent on reducing the damage to the national economy than protecting its citizens. CRMO volunteers thus saw themselves as taking control of their own protection by collecting radiation data.

While the idea of self-monitoring and self-help connotes empowerment, it falls short of this promise when the ability to conduct self-monitoring is dependent on one's own resources. For instance, many CRMOs charge a small fee per sample for measurement because it is necessary to pay for rent for offices and supplies for the detectors. Measurement also requires purchasing a relatively large amount of food (more than a pound), which then is not usable for consumption in many cases. The resource gap also pertains to availability of time and care/wage-earning responsibilities. Some people may feel too stretched to take on another nonremunerated volunteer task. This is particularly the case for women, who often shoulder childcare and elder care—a significant issue in the affected areas, where the relative ratio of seniors in the population increased because younger people tended to evacuate.

Furthermore, more information about contamination might be frustrating rather than empowering when there is a limit to one's ability to act on it. In the case of food contamination, it is impossible for individuals to measure every single item of food that they eat, as each measurement takes time and money. In the case of airborne radiation, one might gain a good understanding of hot spots in the neighborhood and try to decontaminate or avoid going there, but when a strong wind blows up dust that includes radioactive materials, one's ability to avoid it is severely limited.

Many CRMOs are conscious of these tensions relating to volunteerism. For instance, to address the problem of different abilities to access measurement services, many CRMOs made concerted efforts to lower the cost of measurements and often have "open house days," where people can measure their items of concern for free. In addition, most citizen science projects after Fukushima share measurement results on the web and through other means, therefore alerting the public about trends (for instance, certain kinds of fish and mushrooms tend to be contaminated) that would be helpful to those people who might not have time and resources to measure their own foods. There are citizen science groups that consciously monitor areas that are not covered by existing data—even when these zones fall outside of the volunteers' own neighborhoods. For instance, a group in Tokyo is visiting areas in other prefectures to identify hot spots even though their direct constituents might not benefit from such monitoring.

And most importantly, some organizations help people pursue both collective and individualized actions. Some CRMOs, for instance, host lectures not only on the 2011 accident responses by the nuclear village but on the long-standing lack of government and corporate accountability in nuclear governance and the overall pronuclear policies. Information on petitions, elections, legal challenges, and demonstrations is also shared in the network of people involved in CRMOs. These actions that aim to address systemic issues are some of the ways that citizen scientists might circumvent the powerful yet limiting narrative of the individual rights to know and to self-protection.

Voluntary radiation monitoring can provide information that is customized to the needs of individuals in ways that are independent of the authorities. The flip side of this is that there will be people whose information needs are left unfulfilled and systemic issues that are difficult to address. Voluntary monitoring can support both individual self-protection and advocacy for universal protection.

TAKING A STAND

Collective actions to address systemic and structural issues may be necessary, but in certain contexts, they may be seen as too politicized. Engagement in activism has been a tricky terrain for citizen scientists after the Fukushima accident, and some avoided it completely. For instance, neither Safecast nor the ocean monitoring groups participated in antinuclear advocacy or activism. Ocean monitoring was designed as a scientific research project, one of the academic scientists' several research prongs. Safecast started as a "citizens' project" but did not see itself as an advocacy network and avoided taking a side. As Akiko Hemmi and Ian Graham wrote about Safecast, "The commitment to open data led Safecast to avoid interpreting the data that they collected which in turn reduced the risk of them being seen as an anti-nuclear activist organization."[21] Safecast insisted that its commitment was to "open data" and tried to avoid being seen as political. Its mantra is "pro data"—meant to transcend the politicized contestation between "antinuclear" or "pronuclear."[22]

A similar insistence on nonaffiliation with antinuclear movements was present in many, but not all, CRMOs. CRMOs were much more diverse, and the willingness to openly work on an antinuclear agenda and to collaborate with antinuclear activists varied significantly. However, in some cases, efforts to assert boundaries between activism and science were clear. For instance, in one interview that Kimura conducted, a CRMO leader phrased her fear of being stigmatized as an activist as the worry that the group "would be seen as people who are on that side" (*acchigawa no hito*). She went on to say, "We did not want to be activists [*katsudōka*]. There is a woman here who has been issuing newsletters on social problems. People see her as strange. She is a professional activist and not a regular person. We need to look for a new way of social movement. We don't want to be called professional activists [*undō no puro*]. We are regular mothers."[23]

A woman from a different CRMO invoked a similar distinction when Kimura asked her about their activism:

WOMAN: We don't do activism. The original policy of the CRMO is neutrality, not to do something political . . . people who are involved here, personally, they might be antinuclear, but we don't intend to do activism through [the CRMO].

KIMURA: But there are other CRMOs that are antinuclear?

WOMAN: I think they are different. I think the difference is whether CRMOs are "tainted" [*iroga tsuiteiru*] or not. . . . I don't want to taint our measurement results.

KIMURA: Does that mean that you are concerned about how the results might be looked at by others?

WOMAN: Yes. Of course, the others [other CRMOs] are using detectors, so it's not like they can change [the result]. . . . But my feeling is that I don't want to have a predetermined result.[24]

Note here that the questions about activism led this citizen scientist to talk about the idea of "tainting" results. When science provided crucial legitimation for citizens' concerns in the context of the harsh policing of fūhyōhigai, they did not want to risk appearing to politicize science. Activism was potentially contaminating of the legitimacy of the measurement results, and people did not want to be seen as "activists."

Japanese social movements went through tumultuous decades in the 1960s and 1970s, when radicalized factions resorted to violence and deadly infighting. The history of violence was used by the authorities to paint negative images of activism in general.[25] This was part of the reason citizen scientists were careful to separate their work from activism that was seen as too extremist.

Currently in Japan, antinuclear movements are seeking rapid nuclear phaseout, and the victims are mobilizing for compensation. Some CRMOs do collaborate with these campaigns, but others choose more low-profile actions like talking to local government officials and bureaucrats in an effort to change some aspects of radiation protection policies. Pursuing accountability of the nuclear village for the accident, preventing future accidents and contamination, and challenging the dominant nuclear energy policy would necessitate broad-based activism. Citizen scientists can be a part of these movements but face a significant dilemma when politics and advocacy are thought to diminish the legitimacy of science.

CONTEXTUALIZING DATA

As antinuclear movements have long pointed out, the problems of nuclear energy are not limited to whether the food, air, soil, or water is contaminated beyond some set of regulatory standards that would pose a known risk to consumers. There are other issues of environmental justice, such as radiation exposure of marginalized workers at the power plants and uranium mines, the concentration of power plants in impoverished rural areas that provide energy for urban areas, impacts of reactors on fisher folks, as well as the lack of a feasible plan for sustainable waste management.[26] Despite the diversity of issues, the institutions of nuclear regulation pressure residents to frame their concerns in scientific parlance, specifically in reference to a set of measurements. In addition, because radiation is invisible and it is difficult to detect its effects on the body at low levels, citizen scientists typically cannot rely on their own senses and get drawn into complex technical debates about measurement. This situation raises an important question: When citizen scientists use the language of mainstream regulatory science, does it restrict the conversation or open the regulatory domain up to include a wider range of issues?

The mainstream scientists and government and corporate officials in the nuclear village tend to use the language of risk and science to delegitimize nonexperts who critique nuclear energy and related policy. For instance, Nakagawa Keiichi, a professor of radiology at Tokyo University, was quoted in a national newspaper as saying,

> There are various risks in society and if you make a mistake in estimating a risk, you will suffer from consequences. For instance, after the 9.11 terrorist attacks, there was an increase in traffic accident deaths in the US. It was because people became fearful of flying and opted for cars. There is no detectable influence from below 100 mSv exposure, but if you think it is big, it will be the same situation like the post 9.11 US. . . . There are anti-nuclear activists in Fukushima fanning the fear of radiation-related illnesses such as thyroid cancer. They can have whatever political belief but they are ultimately making residents unhappy. . . . People need to believe us, the experts, rather than activists who are not even doctors. Otherwise, they make a mistake in judging various risks that are everywhere in society and they will have to suffer the consequences.[27]

Responding to experts who trivialized people's concerns as stemming from scientific deficiency or irrationality, post-Fukushima citizen scientists have had to invest heavily in the provision of accurate scientific data. For instance, some CRMOs have built relationships with experts at academic institutions (whom they can trust and consider outside of the nuclear village). They seek advice from these experts whenever they detect irregular radiation readings. CRMOs also consult with each other and try to hone their measurement skills. Some organizations send samples to institutions with better detectors when they are not sure about a measurement result. In particular, germanium detectors are able to provide much more precise data, a big advantage over the scintillation detectors that most CRMOs have. A national network of CRMOs has also tried to ascertain the measurement capabilities of each organization by sending them a test sample.

Measuring radiation is extremely challenging for multiple reasons. Different factors can affect the results of a radiation measurement. For example, the contamination level of a particular locale will vary by changing the height of measurement from 50 centimeters to 1 meter above the ground, by taking measurements before or after rain, or by taking samples from ditches or an asphalt road, although they might be only several feet apart. Measurement of foods is similarly complex. The duration of the operation of a detector and the amount of material sampled can influence the result of the measurement. Some people are concerned that exposures even below the government's standard might compromise health. Therefore, many CRMOs tried different ways to detect lower concentrations of contamination—for instance, by running the detector for longer periods of time. Commonly used detectors are not capable of measuring materials other than cesium and iodine, so some organizations are pursuing different equipment in order to measure other radionuclides that contaminate food, particularly strontium.

One CRMO in the northeast of Japan has acquired a detector that can measure strontium, as the staff are particularly concerned with the continuing release of contaminated water into the sea and consequent contamination of fisheries products. Strontium measurement is extremely costly, and there are only a handful of institutions (even including academic laboratories) that can do it in the country. The detector costs more than $100,000, and the preparation of a sample requires multiple steps. Unfazed by the challenge, this CRMO called for donations and purchased the equipment, and in collaboration with a retired scientist from a university, they trained staff members to enable consistent measurement. They even came up with a simplified process of preparing a sample. The article was published in the journal *Applied Radiation and Isotopes*, with a retired scientist as a lead author and one of the female CRMO volunteers listed as a coauthor.[28]

Efforts like these are valuable, as precise measurements of radioactive contaminants might expose a reality that the official reports do not recognize. But there are dilemmas to consider. Investing resources and time to get more precise measurements—getting on a data treadmill—may reduce volunteers' availability for other activities, such as political advocacy. Additionally, when discussion focuses on quantifiable data and technical measurements, important issues of justice and fairness that are not presentable in numbers tend to be sidelined. For instance, Safecast portrayed its data on airborne radiation as "raw data"; the organization did not want to get involved in debates over the adequacy of the government policies or accountability for victims. Their "prodata" mantra crystalizes the view that information speaks for itself. However, many problems are not easily demonstrable in numbers without extensive analysis and interpretation. The uneven distribution of costs and benefits of nuclear energy, workers' exposures at power plants and decontamination sites, and the lack of participation in decision-making in nuclear governance are some examples.

Already the government and nuclear village delimit public debates to measurable scientific and economic benefits and risks of nuclear energy.[29] Pushing back on this framing, critics of nuclear power have pointed to layered problems of nuclear energy and governance that may not be captured by commonly used scientific and economic indicators.[30] For instance, Takagi Jinzaburo, himself a nuclear physicist, argued that nuclear energy reflected the larger problem of a consumerist culture and highlighted the costs that are borne by the marginalized segments of society. He and others also questioned the transgenerational accountability for radiation contamination and waste management. He wrote, "How can we say that the technology is productive when it needs 1 million years of strict control of radioactive materials but produces energy only for several decades? Most scientists do not respond to this question."[31] Calling mainstream nuclear science "epicurean science," Takagi insisted on looking at broader problems of nuclear energy than what are usually construed by the prevailing scientific and economic assessments.[32] Other "rogue" scientists such as Koide Hiroaki have also expressed opposition in terms of environmental justice. Koide wrote, "I am opposed to nuclear energy not because I am afraid of being harmed by accidents. . . . Nuclear energy is thoroughly based on the exploitation and repression of others. That aspect is the reason why I am opposed . . . I would like to use my life to face problems in the world and to make a society where we don't have to force someone to bear unwarranted burdens. And when such a society materializes, nuclear energy necessarily will be abolished."[33]

Citizen scientists may pursue precise measurements to expose hidden contamination that can be used to seek corporate and government accountability. Because health impacts of radiation exposure take decades to emerge, getting a good account of contamination right now is important. Different kinds of data gathered by regular people may in fact be valuable in pursuing justice administratively and legally if contamination data are necessary to prove causality with health effects in the future. It may also be the case that citizen scientists without professional credentials are under even more pressure to speak in the language of science to be heard. The dilemma is that by focusing on measurements of certain aspects of nuclear energy, citizen science might inadvertently contribute to sanitizing the broader political, social, and ethical implications.

Citizen scientists can manage to avoid being reductionist and address both scientific and social justice issues. For instance, some CRMOs worked closely with existing antinuclear movements, combining measurements of food contamination with broader policy issues, including lobbying for legislation to assist victims of the accident and opposing the restarting of nuclear reactors. They have also supported antinuclear candidates in local and national elections so as to push for the phaseout of nuclear power. Scientific data collection increases the credibility of citizens in arguing for the need for official monitoring, remedial actions, and prevention of future harms. By contextualizing the data, connecting it to their own lives, citizen scientists can make an important contribution to public decision-making.

SHIFTING SCALES

Radiation can be measured at various scales. One scale issue that has been salient in radiation monitoring is individual versus district. Citizen science projects can produce intimate kinds of data that are fine grained and rooted in localized knowledge of behaviors, environments, and neighborhoods. But such a personal orientation may inadvertently serve to facilitate the broader trend toward individual responsibility for radiation monitoring and management. For instance, if someone is concerned that produce from a home garden might be contaminated, he or she can bring the vegetables to a CRMO to get them measured rather than approximating from the data provided by the government. The soil and airborne radiation monitoring by residents' groups similarly provides data points that are based on a deep understanding of people's daily behaviors. For instance, some groups measured contamination levels by walking along the paths taken by children to identify hot spots in their neighborhoods, which are not captured by broad measurement data at the district level. Other groups have also measured levels at different heights to see if younger children are exposed to higher levels of contamination due to dust from the ground.

Because of the intimate nature of their data, citizen science projects might imply that measurement is a private act—in a way analogous to measuring one's body temperature. Indeed, the radiation measurement data can be personal when the measurement data are only provided to the clients who bring in the samples for testing. For instance, some CRMOs asked the clients' permission before making the measurement data public, while others restricted the dissemination of measurement results to the organizations' members and not to the general public. These decisions were usually driven by the fear of criticism of fūhyōhigai and of harming food producers by damaging the reputation of the particular food item tested. Some CRMOs mentioned the possibility of lawsuits if the measurement data resulted in the decline in the sales of a food item from a particular area. Similar concerns were raised about soil contamination data and the possible decline in real estate value. In these settings, some CRMOs talked about the "privacy" of clients, describing the data as personally belonging to them.

Responding to individual and varying needs of individuals is important for citizen science projects, but it is also crucial to consider whether this emphasis might be moving the scale at which the problem is conceived. It is noteworthy that promoters of nuclear power are now interested in institutionalizing self-monitoring and management of radiation by citizens.[34] Monitoring the release of radioactive materials from nuclear reactors and ensuring citizen safety is usually considered the responsibility of the operators of the reactors and the government agencies tasked with nuclear monitoring, safety, and environmental protection. But as the government and nuclear industries struggled to clarify the disaster response strategies, self-monitoring became an important component to it. The municipal and national governments started to introduce the virtue of citizen self-monitoring into their accident recovery plans. As the vice minister for the environment said, "We would like as many Fukushima people as possible to have personal dosimeters and take protective actions based on the data."[35] The government's reconstruction budget included the measuring program for the purpose of encouraging citizens to measure their own environment for radioactivity so that they modify their daily behaviors to reduce exposure to radiation.[36]

The government has encouraged the use of dosimeters called the "glass badge" and the D-Shuttle in the affected regions. Both are lightweight, portable dosimeters that can be hung around the neck or carried in a pocket. Glass badges are usually rented out for several months to citizens who then return them to the government, which retrieves the data on the accumulated exposure level and informs the user of the results. The D-Shuttle provides a more detailed, hour-by-hour reading that can be accessed by the user. After carrying it for a day, the user inserts it into a digital reader for three seconds. The reader then displays trends in exposure level for the day as well as accumulated exposure levels. In order to identify reasons for exposure, users are encouraged to keep daily activity logs (fig. 4.3).[37]

Personal measurement and management of radiation could be considered a tool to expand citizens' knowledge and empower individuals. In fact, the Nuclear Regulation Authority's report invokes the language of individual decision-making and respect for autonomy when it mentions personal measurement.[38] ETHOS Fukushima, a risk-communication program that is organized by the International Commission of Radiological Protection and supported by the Japanese government, has similarly emphasized radiation self-management as an instance of proactive and self-responsible behavior that can and should be taken by people in the affected areas.[39]

However, it is important to recognize that the motivations for personalizing radiation management came from various directions, not just citizens' desires for more information. Previously, the government used the standard of 1 mSv/year at the district level as the measure of safety, but decontamination to this standard at the district level proved too costly. Additionally, the cost of supporting evacuees was putting pressure on the national budget. Lastly, from the perspective of local governments in Fukushima, depopulation was a huge problem, eroding their tax bases and the sense of community among remaining residents. They wanted people to return. "Radiophobia" was said to be the reason for evacuation, which could be even more harmful to residents in the affected areas than the actual physical consequences of radiation exposure.[40]

The government makes an argument that self-monitoring enables people to lower their individual exposures even when people live in an area contaminated above 1 mSv/year. Radiation at that level used to be considered too dangerous for regular habitation, making residents eligible for evacuation. To be sure, many

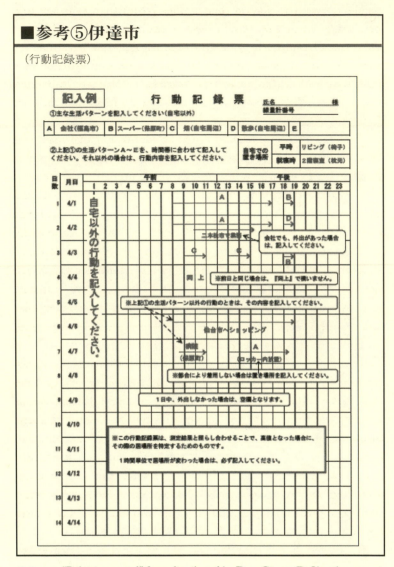

FIG. 4.3. "Behavior record" form distributed by Date City to D-Shuttle users. The form asks the user to identify where he or she was (e.g., office, supermarket, garden, neighborhood) and what he or she was doing at different times of the day. Photograph taken from the website http://www.meti.go.jp/earthquake/nuclear/kinkyu/committee/advisor/2015/pdf/0826_01e.pdf.

people do want to stay in their homes rather than evacuate to a new place where they would have to look for new housing and arrange for changes in jobs and schools. But the government's political and economic calculation also shapes the push for self-monitoring and radiation self-management. Critics have pointed out that dependence on individual dosimeter data entails the possibility of a significant underestimation of actual exposure. Even the manufacturer of the glass badge admitted that the dosimeter was meant for workers in radiological occupations and was not suitable for measuring the public's exposure level.[41]

Citizen science's emphasis on personalized monitoring can be complicit in this broader dynamic of making people personally responsible for their own environmental and health protection. Making citizens responsible for managing their own exposures also normalizes contamination as an always existing condition.[42] Citizen scientists therefore need to consider the balance between the benefits of individuated data versus aggregated data and reflect on how their practices may be shifting the responsibility for monitoring and managing environmental contaminants.

In addition to the dilemma of personal versus systemic scales, the issue of scale is also manifest in terms of the passage of time. What is the appropriate temporal scale of post-Fukushima citizen science? The citizen science projects after the Fukushima accident can be considered a disaster response, which helped residents figure out the level of danger in an emergency situation. The concept of disaster/crisis response may help citizen science by attracting many volunteers, supporters, and funders; however, addressing the root causes of the disaster may require efforts on a longer time scale.

In analyzing the social impacts of disasters, theorists have discussed how natural disasters result in the emergence of a "therapeutic community" that shows a strong social cohesion and serves to ameliorate the disruption caused by the disaster.[43] The triple disasters of the earthquake, tsunami, and the nuclear accident in 2011 certainly brought many people together who felt compelled to do something in a time of collective hardship. Funding from various foundations, private companies, and individuals also became available under the banner of disaster relief and response. Yet studies on disasters have pointed out that disasters and vulnerability to them are rarely abrupt and that there are structural and historic drivers.[44] The lack of transparency in nuclear policy making and the collusion among the industry, the policy elites, and scientific experts, for instance, are systemic problems that plague the overall system of nuclear power in Japan and have a long history that predate the Fukushima accident.

Citizen science projects also have to be prepared to face the long-term consequences of nuclear power. The impacts of cesium from the Fukushima accident will not go away soon, given the long half-life of radioactive materials. Ongoing troubles at the Fukushima Dai-Ichi Nuclear Power Stations and its decommissioning as well as the continuing existence of nuclear reactors and radioactive waste also raise questions about the possibilities of the future release of radioactive materials into the environment.

When citizen science groups emerge in response to a disaster, it is understandable that they focus on the immediate need for data to aid survival and well-being. The disaster frame also helps mobilize the constituency with its emotive appeal of human tragedy. Citizen scientists need to be conscious of the short timeline that commonly accompanies a disaster frame and be ready to assert and address the historic and systemic roots. The consideration of temporal scales can entail a gradual shift from the disaster response framework to prevention, although the latter inevitably involves more politicized questions that challenge the existing power structure.

Conclusion

The vast majority of people who became citizen scientists after the Fukushima accident were nonexperts in the field of radiation measurement, but they were able to master the techniques to measure food, soil, air, and ocean water, sometimes with the help of credentialed scientists. In a highly uncertain environment where the degree of contamination was unclear, citizen scientists strove to fulfill acute information gaps. They provided fine-grained information that was responsive to people's varying needs. Because many people felt that the accident revealed the secrecy and unaccountability of the utility industry, regulatory agencies, and affiliated scientists, they welcomed data from citizen scientists as more objective and trustworthy.

Citizen science projects had diverse structures and outcomes. In ocean monitoring, credentialed scientists took tight control over the design of the project; in contrast, regular citizens organized CRMOs, albeit with some help from sympathetic experts. In addition to the provision of scientific measurement data, some projects resulted in academic publications. Safecast improved measurement equipment and created an online platform for sharing radiation-related information. Many CRMOs provided a space for regular people to network and learn more about radiation and nuclear energy.

The cases described in this chapter also brought up various dilemmas of citizen science. Citizen science can help nonscientists legitimate their concerns, particularly when contaminants are invisible, like radiation. When powerful actors portray environmental and health complaints as irrational and harmful, citizen science helps people present grievances in the language of science. However, it can also inhibit necessary political actions when "good science" is understood to be separate from activism and politics. Another dilemma is that when citizen science projects place an emphasis on the personal nature of measurement data, they risk legitimating the notion that pollution monitoring and management are individual responsibilities.

This chapter also pointed to how citizen science can take on the character of emergency response in a disaster situation to aid in the survival of victims and individual self-help. The idea of crisis spurs the creation of citizen science projects, attracting volunteers, supporters, and funders. But framing citizen science projects as disaster mitigation and the fulfillment of temporary information gaps can obscure the need to address the historic causes and compounding factors of a disaster. These scales may change throughout the lifetime of a citizen science project. Citizen scientists can make the strategic choice to move flexibly between different scales and time frames; for instance, postcrisis monitoring can be used to build interest in longer-term projects to investigate and challenge environmental injustices on a wider geographic scale.

The post-Fukushima citizen science projects may be unique in that they responded to a crisis that was caused by natural disasters, the impacts of which were uncertain and highly contested. Citizen science projects in Japan may also seem very different from the ones in Euro-American contexts, with very different histories of social movements, environmental and energy governance, and the structure of scientific institutions. Without diminishing the importance of attention to such specificity, we observed that citizen scientists in Japan faced some of the same dilemmas that citizen scientists grappled with in the case of fracking. For instance, volunteer-driven monitoring of both radiation and fracking bring up resource gaps and regulatory complacency. Citizen scientists in both cases also struggled to balance activism and science.

Citizen science emerges and grows in divergent historical and contemporary political contexts; hence dilemmas do not lend themselves to a uniform solution. This chapter's examples in this particular sector (nuclear energy) at a particular historic juncture (the Fukushima nuclear accident and its aftermath) do not offer a universally applicable guide to citizen science, but they do illuminate the need for sensitivity to dilemmas that emerge in multiple forms.

5

Tracking Genetically Engineered Crops

In 1998, U.S. regulatory authorities approved StarLink corn, a genetically engineered (GE) crop. However, regulators were concerned that StarLink corn might cause allergic reactions in humans, so permission was limited to animal consumption. Two years later, a coalition of seven environmental organizations, including Friends of the Earth, the Institute for Agriculture and Trade Policy, and Pesticide Action Network, announced that they had conducted genetic tests that revealed StarLink corn in taco shells that were manufactured by Kraft Foods. Further investigation revealed that the taco shells had been made from corn flour from a mill in Texas, which obtained corn from grain elevators that did not segregate GE and conventional grains. There were subsequent discoveries in Japan and South Korea, and in later years, activists found StarLink in Mexico as well.[1]

The effects of the StarLink discovery were dramatic. For months, headlines read, "Taco Bell Recalls Shells That Used Bioengineered Corn" (*Los Angeles Times*, September 23, 2000), "Safeway Recalls Its Taco Shells" (*CNN*, October 1, 2000), "Corn Woes Prompt Kellogg to Shut Down Plant" (*Washington Post*, October 21, 2000), and "GMOs Are Found in Morningstar Farms Products" (*Los Angeles Times*, March 8, 2001). Aventis Corporation,[2] the biotechnology firm that developed the seeds, was forced to buy back grain from farmers and faced costly lawsuits that continued for many years. The incident severely depressed U.S. corn prices for months. Even today, the StarLink case remains emblematic of the unique challenges of introducing living GE organisms into the environment. It set new legal precedents, and it is discussed in hundreds of law and policy research articles. It provided support to efforts to restrict GE seeds and foods in the European Union and other places around the world.

Activists took the lead in discovering that StarLink corn had contaminated food. Today, there are many more examples of citizen science dealing with GE crops. Around the world, people have been testing food and plants to see whether GE material is present. There are other research approaches as well, such as monitoring species affected by the use of agrichemicals linked to GE seeds and collaboratively reviewing published health studies. The diversity of citizen science projects reflects the complex set of concerns that animate the worldwide debate about the biotechnology industry.

Humans have been breeding plants and animals for agriculture for thousands of years, but only in the past few decades has it been possible to make targeted changes to the genetic structure of organisms through genetic engineering. GE plants, microorganisms, and animals are often called genetically modified organisms (GMOs). By inserting specific pieces of DNA, it is possible to add new traits to an organism—for example, making corn plants produce a substance that kills insect pests, making soybean plants resistant to herbicides, or giving papaya trees protection against a virus. GE farm animals, though not in common use today, could produce useful medicinal substances, resist illness, or digest food more efficiently. These innovations have been controversial. Critics of genetic engineering raise questions about the ecological impacts of releasing organisms with novel traits into the natural environment. In addition, food safety and consumer protection activists have speculated that there could be unknown health risks related to eating GE foods.

Controversies over GE crops are not simply scientific disputes; they are multifaceted struggles over the future of food and farming. In the popular media, opponents of GE crops are often represented as fearing unknown "Frankenfoods," but many critiques of GE crops are actually linked to demands for *food justice*—the

idea that "the benefits and risks of how food is grown and processed, transported, distributed, and consumed [should be] shared
equitably."[3] A food justice critique of GE crops points to the ways
that the biotechnology industry perpetuates social inequalities
and environmental problems throughout the modern food system. A related concept is *food sovereignty*, an idea that highlights
control over food production and consumption in the context of
globalization. Food justice and food sovereignty advocates are particularly critical of large corporations—including biotechnology
companies—that have an outsized effect on the food system.

The vast majority of GE plants grown around the world are
engineered and sold by a small number of multinational agricultural biotechnology (ag-biotech) companies. The consolidation
of the industry has accelerated since the time of the StarLink
controversy. Since 2016, Bayer purchased Monsanto, Dow merged
with DuPont, and China National Chemical Corp bought Syn-
genta. The special traits of the plants they sell—mainly herbicide
resistance and the ability to generate a pesticide—are patented by
these companies. Patents prohibit farmers from saving and planting seeds from these plants, which ensures that farmers will buy
new seeds each year. Critics argue that this is an unfair economic
burden on farmers and that this system has a negative effect on
the genetic diversity of agricultural crops. Additionally, herbicide-
resistant plants, which enable farmers to clear weeds from a field
without harming the crop, have dramatically increased the use of
particular herbicides that are manufactured by these companies.
Critics are concerned not only about the health and environmental
impacts of herbicide use but also about the near monopoly enjoyed
by biotechnology companies that control both seed and agrochemicals. In particular, advocates of small-scale farming criticize the
industry for driving up input prices, increasing the indebtedness of
farmers around the world.[4]

How do citizen science projects address issues of power
relations, particularly the dominance of large multinational corporations in the food system? How can citizen science support more
equitable and sustainable food production practices? In what ways

do genetic testing and other forms of citizen science empower people to get involved in decision-making and agenda setting? This chapter probes these questions, focusing on an activist-led study that took place in Mexico during a controversial debate about trade in GE corn. We will make comparisons with an assortment of additional examples, including a study involving the impacts of herbicide-tolerant crops on butterflies, citizen tracking of GE rapeseed (canola) in Japan, and two projects led by organizations that tend to advocate for GE crops. All these projects use participatory research to explore questions about introducing GE crops into food and agriculture systems. These examples illustrate several virtues; for instance, data collected by volunteers can be used in published scientific research, and participation in experiments can teach scientific reasoning.

The Mexican case deserves particular attention because it illustrates how citizen science can strengthen broader movements for social change and justice in the food system. Citizen science can produce knowledge from vantage points that are usually excluded, give voice to the marginalized, and strengthen advocacy for change. In the case of the Mexican corn study, activists concerned with the expanding use of GE seeds were able to answer scientific questions that otherwise were being ignored. By highlighting the industry structure and regulatory system that made food contamination difficult to avoid, the activist researchers began to insert questions of justice into the public debate about GE food safety. And by working together to publicize the findings, the project created a platform for political advocacy for stricter controls on GE seeds and changes to trade policy with the United States.

As in the case of fracking and nuclear energy, citizen scientists tackling questions about GE crops encounter dilemmas as they seek to contextualize the information that they gather. Many projects suffer from reductionism—allowing fairly simple scientific measurements to stand in for extremely complex questions about the food system. Presenting concerns about biotechnology in focused scientific language can be an effective way to get the attention of regulatory officials and the scientific community, yet it can come at the cost of sidelining questions of culture and values. Deciding how to navigate this dilemma is just one of several difficult choices that citizen scientists dealing with GE crops must make.

Detecting GE Corn in Mexico

Soon after the StarLink scandal broke, unapproved varieties of GE corn appeared, somewhat mysteriously, in Mexican cornfields where traditional farming methods were practiced.[5] A coalition of indigenous peoples, farmers' unions, peasant corn growers, and environmentalists carried out a wide-ranging genetic study of corn plants and used the findings to support a growing movement in defense of rural livelihoods and sustainable foodways. What makes this case stand out are the ways that collaborative research was incorporated into a holistic analysis of Mexican corn production, making connections to broader food justice issues, including unfair trade agreements and the oppression of indigenous peoples.

Corn is highly important to the Mexican economy, diet, and culture. It is the staple of the Mexican diet and also holds symbolic and spiritual significance for many indigenous peoples.[6] Additionally, locally adapted varieties of corn are globally valued as a source of agricultural genetic diversity; for this reason, Mexico is home to a network of international corn research centers. Although it is not known what effects, if any, introducing GE corn to Mexico might have on the existing genetic diversity of corn and its wild relatives, both environmentalists and plant scientists have called for caution.[7] Responding to these concerns, the Mexican government has tried to protect native biodiversity by prohibiting the cultivation of GE corn, except for experimental purposes (fig. 5.1).

In Mexico, corn production employs about three million farmers and uses 60 percent of cultivated land.[8] Of those corn producers, an estimated two million farmers follow traditional practices, such as planting multiple maize varieties to suit different agronomic and culinary needs, saving seed from one season to the next, trading seeds with other farmers and communities, mixing

FIG. 5.1. "How to Keep Our Maize Pure." Brochure produced by SEMARNAT, the Secretariat of Environment and Natural Resources, Mexico's environment ministry, around 2005. The infographic explains that people should keep planting seeds from the corn they cultivate in their own fields but avoid planting imported corn that is sold for human or animal consumption.

seeds of different origins, and adapting improved varieties to local conditions. In 2002, corn was produced in these traditional ways on about 5.8 million hectares—80 percent of corn production land in Mexico.[9]

However, the Mexican corn economy has undergone major changes since the end of the 1980s. Under pressure from the World Bank, the Mexican government reduced its subsidization of corn production, transforming the country's food and agriculture strategy from self-sufficiency to import dependency.[10] The beginning of the North American Free Trade Agreement (NAFTA) in 1994 further committed Mexico to replacing domestically produced corn with cheaper imports from the United States. Predictably, this had devastating economic impacts on corn producers, increasing rural poverty and leading to migration from the countryside to the cities and the United States.[11] Not only was U.S. corn cheaper due to government subsidies and economies of scale; it was also mostly GE. By 2000, U.S. corn acreage was already 25 percent GE, mostly insect resistant varieties called Bt, the type of GE corn that produces its own pesticide.

Rural Mexicans and their urban allies resisted these policy changes. Among the most militant responses was the armed Zapatista uprising in 1994, timed with the beginning of NAFTA. The Zapatistas (named after the Mexican revolutionary leader, Emiliano Zapata), brought worldwide attention to the struggle of indigenous people in Mexico to claim rights to land, livelihoods, and political autonomy. Many other *campesino* (peasant farmer) organizations also protested, some joining the antiglobalization movement and the international Via Campesina network, which advocates for food sovereignty and the rights of small-scale producers.[12] In addition, environmental activists became increasingly concerned about GE corn imports from the United States. In early 1999, Greenpeace Mexico activists took samples gathered from ships bringing corn from the United States to Mexico and sent them to a government laboratory in Vienna. The corn shipments were found to contain Bt varieties. Greenpeace Mexico announced the findings in the press and held demonstrations at the port of Veracruz, raising the visibility of public concerns about GE corn.[13]

Professional scientists were also attempting to trace the movements of GE corn, but controversy surrounding those studies generated even greater desire for an activist-led investigation. Ignacio Chapela, a scientist at the University of California, Berkeley, began helping an indigenous village in rural Mexico set up a laboratory to do polymerase chain reaction (PCR) analyses—a method of detecting GE material. Chapela had a long-standing relationship with a small organization that was aiding the community with sustainable development projects, and he intended to provide them with a means of certifying their own corn as non-GE. In late 2000, he and his collaborators were stunned to discover that samples of corn grown in the village already contained traces of GE material. Chapela and his graduate student continued to analyze corn samples, soon publishing their discovery in the prestigious scientific journal *Nature*.[14] The research suggested that corn from the United States had been planted in Mexico (perhaps unintentionally by corn producers using imported "food" grain as seed). The evidence suggested what plant scientists call "transgene flow," the movement of engineered genes into a plant population through pollination.

Greenpeace Mexico and other environmental and food activists responded with alarm to what they viewed as both food contamination and environmental pollution. They submitted a public complaint to the country's environmental regulatory agency, calling for a ban on U.S. corn imports. However, the study was quickly attacked by critics—some working in public relations for the biotechnology industry—who claimed that the research contained serious errors.[15] In an unprecedented move, the editors of *Nature* decided to publish a retraction, agreeing with critics that parts of the research were flawed. However, in 2002, scientists in the Mexican government announced that they possessed data to confirm Quist and Chapela's findings.[16] *Nature* did not accept their report for publication, though, and the results were never published in any other peer-reviewed journal.

In this context of scientific ambiguity, activists with several environmental and antiglobalization organizations in Mexico City decided to facilitate greater public debate about the subject. They sought the participation of campesinos and indigenous people, knowing they would be the most affected by changes to corn production. In January 2002, more than four hundred people attended the Meeting in Defense of Maize, where the findings of contaminated corn were shared and discussed. At the forum, small-scale producers from many regions of the country said that they wanted to know whether their corn was contaminated. This began a project of independent genetic testing coordinated by organizations based in Mexico City but reliant upon the efforts of corn producers to provide samples of their crops. The organizers included the Center for Studies for Change in Rural Mexico (CECCAM, which took the lead on much of the scientific analysis); the Action Group on Erosion, Technology, and Concentration (ETC Group); the National Center for Assistance to Indigenous Missions (CENAMI); and the Center for Social Analysis, Information, and Popular Training (CASIFOP).[17]

Because the organizers aimed to generate a participatory process, they chose villages where it seemed that the study would lead to ongoing grassroots action. Education about genetic engineering was a key part of the project because little was known about biotechnology in rural communities, where education levels are low (fig. 5.2). One of the study coordinators explained, "It's not very easy [to explain] because it is something that can't be seen, and the campesino and indigenous experience has always been with things that can be seen and touched."[18] Despite gaps in molecular knowledge, study coordinators treated corn producers with deference, regarding them as the true "corn experts" because of their superior experience and knowledge of breeding and cultivation.

Coordinators and rural participants took samples from two thousand corn plants in 138 communities across Mexico. The samples were analyzed using commercial kits, manufactured by Agdia Testing Services, which detect proteins associated with particular commercially available transgenic crops. Some of the initial assays were done in the communities themselves with the technical assistance of academic biologists. Later they hired a commercial laboratory in Mexico to do the tests, using the same kits.[19]

The results revealed transgenes in plants growing in thirty-three communities spread across nine states. They detected genetic material from corn plants that were developed and patented by Monsanto, Novartis, and other multinational ag-biotech companies—including StarLink genes. These findings were shared in a five-page press release and press conference in October 2003 and were reported in at least two major Mexico City newspapers as well as Spanish-language newswires. Participating rural communities received posters summarizing the results, and the news was

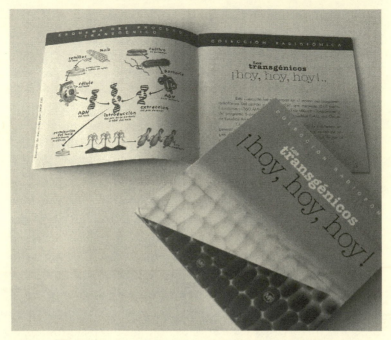

FIG. 5.2. "Transgenic crops, today, today, today!" Title of one of many popular education efforts developed by nongovernmental organizations to inform people about GE corn. Grupo de Estudios Ambientales (Environmental Studies Group), an organization dedicated to sustainable agriculture, recorded a series of audio programs about GE crops to be played on rural radio stations. The CD included a pamphlet with information explaining the genetic modification of corn, including this illustration, originally created for the Sin Maíz, No Hay País (Without Corn, There Is No Country) exposition at the National Museum of Popular Cultures in Mexico City in 2003. Photograph by Avner Ben-Natan.

circulated internationally through an "open letter" to the Mexican government.[20] For an activist-led research project, it was remarkably well publicized.

The corn study proved to be a powerful organizing tool. Investigating cornfields for GE crops became the basis for a new set of solidarities among NGOs, activist groups, and widely dispersed rural communities, which began calling themselves the Network in Defense of Maize. This was a time of growing dissatisfaction with the government's approach to agricultural policy. There were massive campesino demonstrations in early 2003, with protesters demanding that the government reestablish rural support systems that had been dismantled. Participants in the corn study were among the protesters showing solidarity with campesinos and indigenous communities. Believing that the health and diversity of native corn is key to rural communities' well-being, they called on the government to pass a law that would protect native corn varieties. Protests throughout the next several years emphasized the social, economic, and cultural importance of maize and called for a halt to corn imports from the United States. Beyond public protests, participants in the network encouraged producers and consumers to embrace traditional varieties of corn, practice local plant breeding and seed exchange, and refuse to circulate corn of unknown origin. And crucially, as a result of strong mobilization by the Network in Defense of Maize, Mexico still does not allow commercial planting of GE corn, citing concerns about protecting biodiversity.[21]

Other Citizen Science Responses to GE Crops

Beyond StarLink testing and the Network in Defense of Maize, there are many different kinds of citizen science projects dealing with GE crops. In Japan, citizens who are concerned about the spillage of imported GE rapeseed (canola) have been testing for GE genes in plants growing wild near the ports.[22] Each year, Japan imports about 2.4 million tons of rapeseed, mostly from Canada, and more than 90 percent of it is GE. The seeds are shipped by tankers, stored in silos on the ports, and then processed into oil and livestock feed at other facilities. Because rapeseed is very small (about 1 mm in diameter) and the covers on the trucks are not tightly sealed, rapeseed plants proliferate along highways between ports and processors. The concern has been that the spilled rapeseed might cross with the wild and domesticated rapeseed plants grown in Japan and threaten biodiversity in the country.

A group in the western part of Japan called the Non GM Association Aichi started the volunteer collection of rapeseed samples in 2004 and continues the practice to date. Each year, 60 to 80 volunteers go to main roads linking two major ports to processing factories. They pull rapeseed plants that are growing in the wild along the roads and use a testing kit that can identify the presence of proteins found in the herbicide-tolerant plants manufactured by Monsanto and Bayer. The test is relatively straightforward; the crushed sample is screened with two test strips that indicate whether the protein from the GE genes can be identified in blue lines on the strips (fig. 5.3). The kit is imported from the United States and the same kit is used by the Japanese Ministry of Agriculture for their screening. If there is doubt, the sample can be sent to a lab for more thorough testing. Each year, they find rapeseed that has one or both of these genes. For instance, in 2017, 61 volunteers pulled 819 rapeseed plants, of which 78 contained herbicide-tolerance genes.

Several groups are conducting genetic testing of rapeseed across Japan, and the project has become an important tool for solidarity building. Participants include consumer cooperatives such as Seikatsu Club Consumer Cooperatives, Co-op Shizenha, and Green Co-op. Japan Family Farmers Movement, an organization of small farmers, also participates in the network. Citizen science is one of many more joint projects on which these groups collaborate. Additionally, the project is an important mechanism of educating members of these groups about GE issues. For instance, Seikatsu Club itself is a network of thirty-three regional co-ops, amassing the membership of more than 350,000 people. Conducting and communicating about the rapeseed projects is an important mode of membership engagement and education.

A very different response to GE crops can be seen in the Monarch Larva Monitoring Project, which trains volunteers across the United States and Canada to monitor and record monarch butterfly populations. Volunteers identify milkweed, the primary food for monarch larvae, and count the number of larvae and eggs on each milkweed weekly. These data help research scientists analyze why monarch populations are declining. This research intersects with questions about GE crops

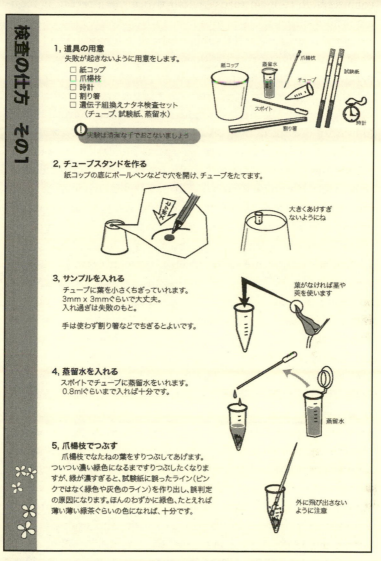

1, 道具の用意
失敗が起きないように用意をします。
☐ 紙コップ
☐ 爪楊枝
☐ 時計
☐ 割り箸
☐ 遺伝子組換えナタネ検査セット
　（チューブ、試験紙、蒸留水）

紙コップ　　蒸留水　　爪楊枝
　　　　　　　　　　　チューブ　　試験紙
スポイト
割り箸　　　　　　　　時計

実験は清潔な手でおこないましょう

2, チューブスタンドを作る
紙コップの底にボールペンなどで穴を開け、チューブをたてます。

大きくあけすぎ
ないようにね

3, サンプルを入れる
チューブに葉を小さくちぎっていれます。
3mm × 3mmぐらいで大丈夫。
入れ過ぎは失敗のもと。

手は使わず割り箸などでちぎるとよいです。

葉がなければ茎や
莢を使います

4, 蒸留水を入れる
スポイトでチューブに蒸留水をいれます。
0.8mlぐらいまで入れば十分です。

蒸留水

5, 爪楊枝でつぶす
爪楊枝でなたねの葉をすりつぶしてあげます。
ついつい濃い緑色になるまですりつぶしたくなりま
すが、緑が濃すぎると、試験紙に誤ったライン（ピン
クではなく緑色や灰色のライン）を作り出し、誤判定
の原因になります。ほんのわずかに緑色、たとえれば
薄い薄い緑茶ぐらいの色になれば、十分です。

外に飛び出さない
ように注意

FIG. 5.3. Instructions for testing seeds for transgenic material. This diagram appears in the manual for the GE testing kit used by the Japan Family Farmers Movement Food Research Laboratory. It is accessible at http://earlybirds.ddo.jp/bunseki/.

because herbicide-tolerant GE crops may be a threat to monarchs' food supply.

Herbicide-tolerant GE crops, such as Roundup Ready Soybeans, are designed to withstand the application of herbicides, particularly glyphosate. They were first commercialized in the mid-1990s, and as a result, glyphosate use has grown dramatically. The use of glyphosate in soybean production increased from 1.4 million kilograms in 1994 to more than 40 million kilograms in 2016; in corn, it grew from 1.8 million kilograms in 2000 to 28.5 million kilograms in 2010.[23] Herbicide-tolerant GE crops enable farmers to kill weeds like milkweed, the only plant on which monarchs will lay eggs. It has been hypothesized that the loss of milkweed from agricultural fields is the reason monarch populations have diminished because they have fewer places to reproduce.

Scientists have used data collected by volunteers for the Monarch Larva Monitoring Project to test this hypothesis, but the data have not lent themselves easily to the question. Citizen scientists make their observations in nonagricultural areas, such as national parks and their own neighborhoods, and year to year they have not seen a decline in the number of butterflies. This is a puzzling discrepancy, since it is well documented that the population of monarchs that migrate to Mexico each year has shrunk. A study by John Pleasants and Karen Oberhauser combined volunteer observations with data they collected in agricultural areas to conclude that the elimination of milkweed in agricultural fields was, indeed, a major contributing factor to the decline in monarch butterflies.[24] In subsequent work, they concluded that "using counts [in nonagricultural areas] can produce misleading conclusions about population size."[25] One lesson to be drawn from this example is that data collected by volunteers cannot replace careful sampling and sophisticated statistical analyses by professional scientists. However, the value of citizen science here should not be discounted; it enabled the researchers to conclude that the decline of milkweed in agricultural fields has not been compensated with a sufficient increase in monarch habitat in nonagricultural sites. The study provides scientific support for efforts to plant milkweeds as a way to increase the size of the monarch population.[26]

Another distinctive example of citizen science dealing with GE crops can be seen in a project initiated by an organization called Biology Fortified Inc. Biology Fortified describes its mission as enhancing "public discussion of biotechnology and other issues in food and agriculture through science-based resources and outreach." The organization is known for its positive stance on GE crops and its cute mascot, a plush GE ear of corn ironically named Frank N. Foode. It is headed by Karl Haro von Mogel, who has a PhD in plant breeding and plant genetics with a minor in life sciences communication. The organization provides consulting services on public outreach on biotechnology, including social media campaigns, curricula for K-12, and citizen science projects.[27]

One of the projects that Biology Fortified devised was a citizen science experiment to test whether squirrels and other wild animals dislike eating GE corn, a claim that has circulated on the internet. As von Mogel described in the blog, the citizen science project was conceived of as public outreach. He wrote, "Research *can also be outreach*. . . . We tend to think of research as discovering knowledge, and outreach as communicating knowledge. I think differently. I think that good research can also be designed as outreach, to both discover and communicate knowledge at the same time. And I believe that we can take this model and apply it to even bigger questions about food, biotechnology, and agriculture. Together, we can change the way that science is done and create a better informed society" (emphasis original).[28]

Using a Kickstarter-like website called Experiment.com, they raised more than $13,000 for easy-to-use experiment kits.[29] The project was also funded by organizations such as the American Society for Plant Biologists. Experiment kits were assembled for about six hundred participants. Two ears of dried corn, one GE and one conventional (both donated by Monsanto), were sent to volunteers, who were instructed to place the corn on a stand outdoors and monitor whether animals were eating either or both ears.

Volunteers submitted their observations, including photographs, to Biology Fortified, where scientists compiled and analyzed the results.[30] However, there have been significant delays in reporting the findings. Donor-participants have been posting comments on Experiment.com since 2016 inquiring about the results of the study. A recent blog post on the Biology Fortified website indicates that the organization severed ties with one of the organizers of the experiment because of his previously undisclosed connections to a biotechnology company, indicating that they would redo his portions of the research. As of early 2019, no results have been published.[31]

Yet another distinctive approach to participatory research was initiated by the Alliance for Science, a program at Cornell University that "seeks to promote access to scientific innovation as a means of enhancing food security, improving environmental sustainability and raising the quality of life globally."[32] The program is funded by the Bill and Melinda Gates Foundation and generally takes a probiotechnology stance. It provides short courses, online materials, and global leadership programs not only in new agricultural technology but also in science communication.

In July 2016, the Alliance for Science put out a call for volunteer researchers to review the abstracts of twelve thousand scientific articles evaluating the safety of foods derived from GE crops. Volunteers were asked to read one hundred to two hundred abstracts to assess whether a given article "makes a determination on whether biotechnology has an implicit, explicit, or neutral statement on human health." In return for their labor, participants would be named in the acknowledgments section of any subsequent published articles.[33] In 2017, the journal *Frontiers in Public Health* published a critique of the study's methodology, noting, among other things, that "conclusions about the safety of a GM food cannot be derived purely from the information provided in the abstract."[34] This is because authors of scientific papers may draw different conclusions about the implications of animal-feeding studies for human health; some may interpret observed changes (e.g., alterations to rat blood biochemistry) as unimportant, while others may view them as a cause for further study and caution. These interpretations, and not the raw data, would be conveyed in the article abstracts. While the results of the Alliance for Science study have not been published (as of early 2019), this may be an interesting example of citizen science to watch because of the unusual and controversial study design.

Clearly, citizen science responses to GE crops have been extremely diverse. Some encourage the participation of people whose perspectives would not normally be represented in scientific work, such as indigenous corn growers or concerned consumers; these may create an opportunity for activism to change public policy about production and trade in GE crops. Others do more subtle political work, emphasizing scientific education and awareness of expert assessments. What can we learn about the merits and drawbacks of these different approaches to citizen science? As in previous chapters, we consider how each project contends with common dilemmas related to volunteerism, tensions between activism and scientific inquiry, the need to contextualize data, and geographic and political scales.

Citizen Science Dilemmas

VOLUNTEERING

In the previous chapters, we raised some concerns about shifting public responsibilities—such as monitoring environmental resources and funding scientific research—onto volunteers. Volunteerism aligns with neoliberal attitudes about public funding for science and environmental oversight; however, the range of possible volunteer roles is wide, and recognizing this diversity is essential to understanding the politics of citizen science today. Citizen science projects around GE crops position volunteers as activists, research assistants, environmental stewards, or recipients of information. The social implications of these positions are often obscured when volunteer opportunities are presented as generically filling knowledge gaps.

Consider the possibility that volunteer monitoring for GE contamination can send the message that government action is unnecessary. For example, volunteer monitoring could reduce the burden on government agencies to carry out biosafety practices. Mexican corn monitoring and Japanese rapeseed monitoring could easily fall into this trap. But Mexican and Japanese organizers never intended to replace regulatory science; in fact, they used citizen science to push for more government oversight. For instance, the Japanese monitoring groups believe that their citizen science projects have encouraged the regulatory agencies to continue their monitoring. In one interview, a key organizer for a group in the west of Japan said, "I am pretty sure that if we stopped doing this (volunteer monitoring), the government would stop monitoring."[35] Every year, the delegates of the networked groups bring their citizen science data to environmental offices at the municipal and national levels to argue for more rigorous monitoring and stricter policies around GE crops.

In the Mexican case, social movement activists initiated the studies with the intention of supporting collective political action and not merely supplementing the scientific data collection. Volunteers were not just helping gather data; they were learning about the issues and working to change public policy and agricultural practices. Furthermore, participants demanded that public agencies invest more in studying the issues that mattered to them. Volunteering to gather corn samples was not just a way to provide important scientific data; it was also a means to challenge the ag-biotech industry and regulatory authorities that did not appear to prioritize corn producers' interests.

In many citizen science projects, volunteers are placed in a subordinate position to experts, limiting their capacity to set the research agenda or determine the uses of the findings. Consider the example of monarch butterfly monitoring described earlier. Some observers of citizen science have critiqued this model of participation because it frames volunteers as research helpers rather than cocreators of a scientific investigation. Yet the style of participation in the research is not the only aspect of volunteering that matters; perhaps more important here is how volunteers engage in practices of citizenship.

Many people feel passionate about monarchs and volunteer to help out of concern for the insects' survival. Yet projects like these could do more to guide volunteers into more active forms of public engagement.[36] Eva Lewandowski and Karen Oberhauser of the Monarch Lab surveyed leaders of twenty-eight butterfly citizen science projects and found that more than 90 percent saw conservation as the broader goal, and 30 percent included information on habitat loss on their websites. They noted the potential for these projects to generate more conservation and advocacy activities: "There remains great potential for projects to offer donation opportunities, directed either to themselves or to related conservation organizations. Similarly, there is much untapped potential for projects to share information on how volunteers can engage in conservation outreach and education by giving public talks or by contacting the media to initiate a news story." The authors went on to "recommend that citizen science projects provide their volunteers with information on donations and outreach opportunities in addition to direct habitat conservation."[37]

In the case of monarch monitoring, volunteers provided data that helped advance research on an important public issue: the role of glyphosate in habitat destruction. Glyphosate use is rising partly because ag-biotech companies are marketing bundled packages of herbicides and GE seeds that can withstand the weed killers.[38] These companies have leveraged their financial and political muscle to subvert actions to regulate glyphosate. Monarch research focusing on the effects of glyphosate remains controversial because some contend that other factors, such as climate change and habitat loss due to logging, have a greater impact on monarchs. Conservation activists have used findings from the monarch study in a campaign to place the butterflies on the endangered species list. Thus this study is a good illustration of how public interest in an environmental issue can be steered toward policy-relevant research that has an impact. But if glyphosate is indeed a major cause of the monarch population decline, reversing it will probably require concerted, broad-based, and political actions that address the entrenched power of multinational ag-biotech companies.

Other organizations have mobilized to challenge the increased use of herbicides and growing corporate control over seed and agrochemicals. For instance, activists supported California's Proposition 65 to officially list glyphosate as a carcinogen and succeeded by writing letters, lobbying, signing petitions, and taking to streets. Non Toxic Neighborhood campaigns are helping communities petition to stop the use of glyphosate in parks and playgrounds in localities.[39] In contrast to these political actions, the way that volunteering is envisioned in butterfly citizen science is not a direct route to participation in these forms of advocacy and collective actions for institutional change. Furthermore, it is possible that volunteering might deter further political action if citizen scientists feel they are already doing their part to save the monarchs.

In some cases, such as the Biology Fortified study about wild animal consumption of GE corn, the only action for volunteers is a simple research task. Biology Fortified is presented as a fun science project, something that kids might undertake with their parents or teachers. Indeed, unlike collaborations between professional scientists and concerned community groups that we have seen in other examples, this project does not appear to have been designed with input from those who expressed concerns or posed questions about wild animal consumption of GE corn. While volunteers may be curious about the topic, nothing in the recruiting material indicates that the organizers are seeking volunteers whose livelihoods or health depend on the outcome. Yet even here, volunteers may have a subtle political role. Citizen science creates an active mode of engagement that can shift people's attitudes and beliefs. When used as "outreach" about a controversial technology, citizen science can be an effective way to encourage people to arrive at the organizers' point of view about GE crops and communicate those ideas more widely.

In previous chapters, we saw that volunteerism came with some dilemmas. Volunteer water monitors struggled with decisions to do work that they believed state regulators should be doing. In addition, those projects—like the citizen radiation measuring organizations (CRMOs) in Japan—resulted in uneven attention to different places. The case of nuclear radiation monitoring raised an additional issue: voluntary monitoring could facilitate the retrenchment of public monitoring and protection of environmental health. This chapter suggests that in addition to these problems, organizers and participants in citizen science also need to consider what kind of political role volunteers will play. Citizen science projects often couch themselves as apolitical and nonpartisan, but every citizen science project is guided by values, even if they are difficult to discern. Citizen science projects may attract volunteers because they seem fun, educational, and altruistic; hence participants may not think about their contributions beyond what is assigned by the project organizers. But the delimitation of volunteer roles can sometimes obfuscate rather than illuminate the broader environmental and political problems. The next section further probes the implications of positioning volunteers as activists for change.

TAKING A STAND

When activists carry out research projects, they make their political agenda explicit. For them, one of the biggest dilemmas arises when opponents attack their studies for being "biased" toward a political outcome. Those who believe in the benefits of biotechnology frequently go on the offensive against those who publicize negative impacts of GE crops, accusing them of being "antiscience" or introducing bias into their scientific assessments. The "antiscience" label is often misplaced and used as a distraction from the cultural, social, and ethical problems that activists are trying to confront. Greenpeace, an environmental organization that was involved in GE monitoring, for instance, has endured such criticisms over the years. More than a hundred Nobel laureates signed a letter accusing Greenpeace of engaging in "opposition based on emotion and dogma contradicted by data" and urging the organization to "recognize the findings of authoritative scientific bodies and regulatory agencies, and abandon their campaign against 'GMOs' in general and Golden Rice in particular."[40] Golden Rice is engineered to produce vitamin A as a way to reduce deficiencies that lead to blindness. Greenpeace contends that the resources spent on genetically engineering rice could be better spent on alternative solutions to malnutrition, emphasizing the problems of industry consolidation and control over seed.[41] Critics try to invalidate Greenpeace's analysis of the corporate-dominated food system by portraying it as "emotion and dogma" rather than science.[42] Yet of course, hunger and malnutrition are political issues, not things that can be resolved by regulatory science or a single technology.

Labeling anti-GE activists as "antiscience" also obscures the many examples where activists are trying to get the science right. The case of wild rapeseed monitoring in Japan illustrates the seriousness with which some anti-GE activists treat their scientific work. The citizen scientists are motivated by the problem of corporate control and dependence on imported food, but they are also interested in understanding the actual status of gene contamination. They do not take science lightly and have tried to increase the credibility of their data through various strategies. For instance, they are using the same test kits that are used by government agencies, including the Ministry of Agriculture. They have also collaborated with the National Institute for Environmental Studies. When data are not clear, they send the sample for a second screening at a professional laboratory. The activists are transparent about both their political positions and their research methods.

In contrast, some examples of citizen science seem to camouflage a political position with the aura of science, hiding unstated assumptions and values. Consider the experiment by Biology Fortified, described earlier, to assess wild animals' preferences for GE or non-GE corn. Its organizers emphasized the scientific neutrality of the project and the scientific credentials of people involved in the organization. They positioned themselves as neutral experts trying to deliver knowledge to laypeople. They highlighted that the project followed the standard scientific protocol; the genetic identity of each ear of corn is not revealed to participants to avoid biasing the observations. This is a good practice for avoiding selective observations. However, the project seems to have a political mission that is not communicated to the volunteers: to dispel fears of genetic engineering by challenging a myth and enabling people to familiarize themselves with rather mundane-looking ears of corn. By focusing on one fairly outlandish myth about GE crops, the Biology Fortified experiment focused attention away from the more contentious aspects of biotechnology corporations and the trend toward highly industrialized food production. As evident in the director's own writings, citizen science is "outreach" by scientists to communicate the benefit of GE crops to regular people. The overall effort serves to promote acceptance of the biotechnology industry.

Although citizen science projects like rapeseed monitoring in Japan have explicit social change goals, it is important to recognize that even seemingly neutral citizen science projects have implicit assumptions and objectives. Is it better to give the illusion of neutrality or to openly pursue participatory research that is motivated by political aims? Portrayed as "antiscience," social movement activists may feel the lure of justifying their positions scientifically, but as we will discuss in the next section, citizen science comes with the risk of overemphasizing the factual, rather than moral, ground from which they make their arguments.

It is also important to recognize that there can be disagreements within activist-led citizen science communities about the best strategies for establishing credibility and gaining a voice in decision-making. This was evident in the case of Mexican corn testing. Within the Mexican scientific community, there was a great deal of concern about GE corn, which created an opportunity for activists to be heard. Some government officials at particular agencies were willing to listen to representatives of activist organizations who took care to establish their scientific credibility. These activists emphasized their use of test kits and laboratories that are also used by professional scientists yet made few political gains through these efforts. The activist researchers communicated their findings in press releases and posters rather than lengthy scientific articles. Professional scientists tended not to take the results seriously, saying that they needed more detail about the methodology before they could do anything with the results. This frustrated some community organizers within the network and led to a schism within the Network in Defense of Maize about whether it is worth trying to adhere to professional research standards.[43]

As organizers of the Mexican corn study discovered, activist leaders of research projects need to be savvy at reading the social situation. Will working with a certified lab head off potential accusations of bias? Are some political institutions more open to activist-led research than others? What are the best ways to be heard in this situation? How should we proceed when officials simply ignore our research? These questions, while difficult to answer, are essential to the practice of citizen science.

CONTEXTUALIZING DATA

The rationalization of the governance of GE crops poses a dilemma for citizen scientists. Critiques of GE crops often situate data in the context of broader problems in the food system, particularly concerning the influence of large corporations. Yet decision-makers such as environmental regulators and science-funding agencies tend to address the question of GE crops in a more circumscribed way, focusing on things that can be measured scientifically. When GE crops were first being introduced in the 1980s and 1990s, many governance bodies judged questions of community and socio-economic implications of technological change to be unrelated to regulatory decisions.[44] Deeply political questions about how we should steward the plant life on which humans depend have been decontextualized, transformed into relatively simple questions about the demonstrable effects of particular plants on environments and human health.

In this context, participating in scientific research can enable concerned people to have a say in decision-making. For instance, in the StarLink case, citizen scientists brought about a recall of the controversial seeds and the contaminated foods because they were able to demonstrate a violation of government regulations. But as we have noted, when citizen science focuses solely on collecting scientific data, the broader politics of the situation can be obscured. If the matter of GE crops is reduced to questions that scientific data can address, where does that leave the complex problems of seed patents, the loss of biocultural diversity in farming practices, and the growing indebtedness of farmers?

In the Mexican maize case, regulatory decisions at the national level have been based mainly on biodiversity concerns—a scientized way to talk about the value of diverse nonhuman life. When GE corn is framed as a threat to biodiversity, the focus is only on ecological preservation, not cultural and economic survival. Thus despite keeping GE corn out of commercial production in Mexico, activists do not believe their demands for food justice have been met. The government's policies indicate continued prioritization of free trade commitments at the expense of rural and indigenous livelihoods. Imports of U.S. corn continue to threaten the livelihoods of Mexican producers and the future of traditional corn farming.

Citizen science projects have the potential to resist overly narrow framings of environmental problems, bringing social and cultural concerns into dialogue with scientific observations. Yet in some examples, citizen science projects seem intentionally designed to rationalize the debate about GE crops in a way that brings the public's questions in line with narrow regulatory criteria. This critique can be made of the Alliance for Science project that recruited volunteers to read scientific abstracts. Participants are not encouraged to view GE crops holistically; they are asked to summarize factual claims made in the abstracts of individual research articles. Ironically, even while the project encourages volunteers to approach the topic with a scientific attitude, it does not fully empower volunteer researchers to participate in scientific inquiry. It is likely that some studies will present data that could be interpreted in multiple ways, only one of which would appear as a conclusion in the abstract. Yet volunteers are expected to report only the conclusions reached by the authors of each article—without interpreting data from their own perspectives. Thus the project discourages healthy skepticism about published research.

Education and outreach are clearly important virtues of citizen science but can have undesirable outcomes if they emphasize a reductionist perspective on a complex cultural and ecological issue. When faced with a policy environment in which issues must be framed as "scientific" in order to be heard, citizen science can give voice to participants; however, an emphasis on measurement often sits in tension with the virtue of contextualizing data with the participants' broader experiences.

SHIFTING SCALES

What is the right scale for citizen science projects to deal with GE crops? Environmental monitoring could reveal local changes in plant/animal populations, but understanding the cause of the changes might require examination of global trade flows and the patterns of pollution and habitat destruction. Conversely, studies at a national scale might reveal large patterns but neglect individual experiences and local values.

In the Mexican case, activists focused on identifying the corn plants that were not approved for cultivation in their country. However, as the Mexican activists realized too well, discovering GE corn shed light on only a fraction of the larger problems: unfair trade agreements, the increasing market share of imported foods, and threats to the traditions and livelihoods of peasant communities. Thus these concerned farmers and indigenous communities not only tested corn plants but also pushed for the ban on corn imports and revitalized practices of local seed exchange. Their collaborative research efforts addressed different scales of the problem—from genetic identification to global political economy.

The same awareness of different scales is also present in the Japanese anti-GE movement and its collaborative research project. Volunteers in this project focus, at one level, on discovering spillage of GE crops and crossbreeding with other species in their localities (fig. 5.4). The spillage can be reduced by better management practices, such cleaning and sealing the trucks used by particular corporations that handle rapeseed in their area. The groups are indeed demanding these measures, but they also target food and trade policies. A bigger and more fundamental problem is Japan's high dependence on imported rapeseed and the disappearance of

FIG. 5.4. Volunteers recording rapeseed in Japan. Photo courtesy of the Non GM Association Aichi.

domestic alternatives. The groups testing rapeseed samples near the ports realize that they must connect the local escapes of GE rapeseed to a national shift in food production and global economic forces. As one citizen scientist in Mie prefecture that Kimura interviewed said, the GE problem in Japan is one piece of a larger "crisis of food and agriculture in Japan."[45]

The "crisis" has many complicated roots, but one factor is trade liberalization in agriculture. For instance, Japanese rapeseed production started to shrink when soybean (another source of cooking oil) import was liberalized in the 1960s due to pressure from the United States. The citizen science group realized that pushing for technical solutions to the GE rapeseed spillage would pressure smaller oil processors but not multinational ones. This was because bigger oil processors tend to have manufacturing factories on the port, so they are not responsible for much of the spillage along the road, while smaller local oil processors are located farther inland. Requiring these smaller local processors to install new equipment to help reduce spillage might further accelerate consolidation of the oil industry in the hands of multinationals. Already domestic production of oil crops has dwindled, and imported rapeseeds outcompete domestic crops.

With awareness of these challenges, the Japanese groups are using the participatory research project as one strategy for their movement to build safe and sustainable food systems in Japan. They work at different scales: they lobby the government and corporations to demand more information disclosure and accountability in their use and sales of GE crops, GE ingredients, and pesticides; engage in international and domestic dialogues on the Cartagena Biosafety Protocol (a United Nations agreement); and participate in the global GMO Free Zone movement. The Japanese anti-GE movement—of which the citizen science groups are a significant part—has been successful in mobilizing consumer opposition. As a result, although many crops have received approvals for farming and human consumption, farmers are growing no major GE crops in the country as of 2018.[46]

Both of these examples feature activists who use participatory research for solidarity and movement building in order to tackle injustices in the food system. Citizen scientists make local observations, providing detailed data that can suggest local conservation actions, while maintaining focus on the broader forces behind the environmental and health problems that concern them. Moving between scales in this way is vital to environmental science and politics today.

Conclusion

The StarLink case popularized a relatively new method for food activists to generate change. Genetic testing of plants and food became part of the strategy for many other advocacy organizations that were concerned about the impacts of GE crops. Repeatedly, activists around the world conducted research to show that supposedly non-GE foods were in fact contaminated with GE products either through blending grains or through genetic mixing in the fields. These actions brought public attention to the difficulty of keeping GE crops separate in the food supply, highlighting instances where unapproved or unlabeled GE material was discovered in food in the market or growing in fields.[47]

Despite all these accomplishments, the case also shows that the outcomes of citizen science can vary, particularly when addressing complex problems and seeking changes at the system level. In the United States, where the StarLink controversy started, public policy remained favorable toward GE crops despite the seemingly scathing blow inflicted by the food study. Aventis did have to pay farmers who suffered reduced sales of corn and also those consumers who claimed they suffered from allergic reactions. But overall, the StarLink scandal did not stop expanding acreage in GE corn. Even food labeling laws did not materialize in the United States. Sales of GE seeds increased, and industry consolidation intensified.[48] From a food policy perspective, very little was changed. Indeed, GE crops remain hotly debated, and citizen science has grown and diversified in response to ongoing public concerns.

Public contestation over GE crops has a wide range of effects on citizen science on this topic. We have seen that citizen science projects vary significantly not only in their research methods but also in the style of engagement with political, cultural, and social issues. Food and agriculture activists in Mexico and Japan have used collaborative research as a method of community building, political engagement, and awareness raising. Rather than taking the GE controversy as simply a technical challenge, these groups postulate that the issue is about the unequal distribution of power and decision-making opportunities. Their prognosis of the problem spans from the technical to the international political economy.

At the same time, we have seen that groups aligned with pro-biotechnology interests have also conducted citizen science projects. These projects cast themselves as neutral and scientific. To the extent that some groups admit their particular orientation, citizen science is framed as "outreach." Yet they seem to steer participants toward greater deference to scientific experts and delimit the discussion of GE crops away from important topics such as industry consolidation, corporate-led trade liberalization, and regulatory capture by a powerful industry. This framing suggests that they regard the public with a "deficit model," a framework that interprets people's opposition to certain technology as rooted in a lack of knowledge. In stark contrast, the coordinators of the Mexican corn study repeatedly emphasized that the participants were the real "maize experts" even if they didn't fully understand biotechnology. This allowed for a more equitable exchange of knowledge and the deepening of social solidarity.

As we've maintained throughout this book, citizen science has tremendous potential to generate knowledge from historically marginalized perspectives, give voice to participants, and shift power inequities. But this is not possible in projects based on a deficit model of the public's understanding of science and technology. Complex sociotechnical problems call for multiple perspectives and forms of knowledge. We should be wary of citizen science that devalues the knowledge and ideas that participants bring to the research and instead seek new models of participation and collaboration that enable multidimensional analyses of the complex problems of our time.

Conclusion

A Vision of Science by and for the People

Citizen science has garnered strong interest in the last several decades. Not only academic scientists but also policy makers and nonprofit organizations are now embracing the concept. Amid all the attention, citizen science has attracted both strong enthusiasts and detractors. Supporters claim citizen science produces a wide range of virtues, such as scientific literacy, engaged citizens, responsible research, democratized science, and stronger environmental policing. Detractors caution that citizen science—at least in some forms—exploits volunteers' free labor and represents a corporate takeover of scientific priorities.

A central goal of this book has been to move beyond generalizations about citizen science by exploring the diverse ways that citizen scientists navigate common dilemmas. Even within the small set of cases we covered in this book—fracking, genetically engineered (GE) crops, and nuclear energy—we saw tremendous diversity. Participatory research efforts differed not only in terms of the issue areas and their contribution to scientific knowledge but also in the style of participation and their connections to policy and political processes. Citizen science projects are situated in different social contexts, reflecting different histories of activism, state-science relations, industry structures, and forms of civil society. Furthermore, these contexts are dynamic and can change over the life of a citizen science project. Recognizing this social complexity, we did not provide a list of "good" and "bad" aspects of citizen science or a typology of citizen science. Rather, our strategy has been to use diverse examples to illuminate the choices that environmental citizen scientists can make.

The dilemmas should be familiar to readers by now, and their interconnections may be evident as well. Citizen science today occurs in a context where public funding for scientific inquiry is diminishing, and resources for regulatory science and enforcement are being stripped away. These changes mean that the significance of *volunteering* and ideas of good citizenship are shifting in both contentious and subtle ways. Some citizen scientists may choose not just to volunteer their service but to *take a stand* for environmental justice. Yet once citizen scientists are seen as activists, opponents are likely to charge that their data are biased and politically motivated. Emphasizing the rationality and scientific basis of arguments can provide credibility but has drawbacks as well. When citizen science emulates professional and regulatory scientific practices, it can reduce participants' attention to the political, cultural, and social factors that are necessary for *contextualizing the data*. Unlike many scientific professionals, citizen scientists have intimate knowledge of local and personal *scales*. Yet deciding how to achieve the greatest impact with citizen science may mean reframing issues as having national, transnational, or even planetary dimensions.

Some might argue that these dilemmas are only applicable to a certain type of citizen science. The cases in this book focused on major national and international controversies involving air, water, soil, and food contamination, where participatory research held the potential to address issues of widespread public concern. Perhaps the stakes are not so high, and dilemmas not so acute, in citizen science on less controversial matters. But we believe there are also difficult choices to make—for example, when monitoring bird or insect populations. Recall the example of monarch butterfly monitoring discussed in chapter 4 and imagine if the organizers of that study had chosen to work with antipesticide activists in addition to butterfly-loving volunteers. Was this a conflict averted or an opportunity lost? And consider how dilemmas related to volunteerism arise when a research topic is not of great public interest.

Scientists and conservationists could devise ways to entice volunteers, or they could choose to advocate to change science funding policies so that volunteer labor is not as necessary.

Some dilemmas may likewise seem unproblematic for social movement–based citizen science, where a commitment to environmental justice may provide a moral compass for action. But consider the community garden activists mentioned at the start of chapter 2. Imagine that instead of striving to quantify the productivity and nutritional benefits of community gardens, they documented their social, cultural, and historical values in stories and artwork. Was their choice a strategic misstep or a savvy response to the rationalized political landscape? Or take the example of Mexican corn activists. Within that social movement, organizations argued over the appropriate scale of problem solving—the local community, the state, the nation, or the international community. The plan for the participatory study had to be flexible enough to serve multiple members of the coalition who were pursuing different strategies. It is not hard to imagine that a different set of choices could have led to greater schism.

For all these reasons, grasping these dilemmas and how they are related to the ever-shifting social landscape is essential to practicing any environmental citizen science today. Citizen science projects require planning—not just about research design and data collection tools but also about strategies for responding to the dilemmas that bedevil participatory research: austerity pressures, presumed boundaries between science and activism, reductionist tendencies that privilege quantifiable scientific and economic indicators, and difficulties moving between scales of environmental problems. There is no "one-size-fits-all" way to confront the political challenges involved in carrying out participatory environmental research, yet we believe that initiators of and participants in these projects can do more to facilitate discussion about these dilemmas.

While each chapter of this book has elaborated the challenges and difficult decisions facing citizen scientists, we remain hopeful about the promise of public participation in scientific research. Our interest in citizen science stems from our desire for a more equitable and just society. From this perspective, we value participatory environmental research that does at least three things. It should bring diverse participants into the research process in order to produce knowledge from vantage points historically left out of science. It should learn from and support efforts to create a more equitable society. And finally, the collective practices of citizen science should construct a platform for broader political participation. These are the hopes for citizen science that we brought on this expedition. Having journeyed through the streams and gas fields of Pennsylvania, postmeltdown Fukushima, Mexican corn fields, and many other situations, how do these ideas hold up?

DIVERSIFYING KNOWLEDGE PRODUCTION

Citizen science has been a tool for subaltern groups to articulate their life experiences and a way to counter dominant narratives spread by more powerful actors. When Japanese mothers, Mexican corn producers, and Pennsylvania watershed protectors bring their unique points of view into the research process, new ways of thinking about environmental change become possible.

Take the example of post-Fukushima radiation monitoring. The Japanese government has encouraged people to monitor their own exposure levels and the contamination of their environment, but in general, the government's monitoring programs have worked to suppress public concerns about radiation exposures. In contrast, the citizen radiation measuring organizations (CRMOs) questioned expert assurances of safety. When the powerful coalition of academics, bureaucrats, the utility industry, and other business sectors insisted that the nuclear accident did not cause significant contamination, volunteer researchers helped citizens critique such claims. Their work ensures that the impacts of the nuclear disaster will not be ignored or forgotten. This example, among others examined in this book, inspires hope for the diversification of knowledge production through citizen science.

In a more general sense, citizen science can provide resistance to assaults on environmental science. As budgets for regulatory science are cut and scientists increasingly rely on corporations and wealthy individuals for research funding, volunteers can help ensure that research continues on environmental topics that do not fit the narrow priorities of funders. The increased role of corporations and wealthy individuals in science funding has raised a profound question about whose views and priorities are driving scientific research. For instance, energy companies, including ExxonMobil, gave Stanford University a grant of more than $200 million to create the Global Climate and Energy Project. At the same time, ExxonMobil spent significant energy and resources to create societal doubts about climate change, including funding conservative think tanks.[1] Corporate and private donations may be celebrated as key to "public-private partnerships" but can also erode the ideal of open public science.

In this context, citizen science seems to provide an important counterweight to the private funding that increasingly shapes research directions. While the voices of a select few corporations and rich individuals are getting louder in science, citizen science can help diversify the perspectives and experiences that inform research. *Citizen*-based science brings the *public* into science, grounded in an idea of science as integral to democratic society and its citizenry.[2]

Furthermore, when cuts in governmental funding create gaps in environmental monitoring and research, citizen science can create knowledge that is needed by ordinary people. In the case of volunteer watershed monitoring, state and federal agencies appear unable or unwilling to rigorously investigate the impacts of fracking in rural communities. Volunteers seek to fill in the gaps and take pride in the knowledge they produce. However, while water monitoring volunteers express satisfaction that they are "doing something" and strengthening ties to their community and ecosystem, many also believe that they are "doing what the state should be doing."

Indeed, citizen science is not a remedy for the privatization of science or shrinking science funding. Furthermore, participatory research does not necessarily or automatically lift up the perspectives of marginalized people. Projects that work intentionally and purposefully with marginalized social groups provide inspiring models, but we have seen numerous citizen science projects that do not allow the volunteers to set the research questions; often, participants simply do data collection on projects designed by experts. Furthermore, the inclusion of marginalized groups has often been sacrificed, as projects rely on volunteers who have discretionary time and are already comfortable working with scientific information. Even as citizen scientists accumulate data that matter to them, uneven participation means that other voices are not yet heard.

Challenging these tendencies will not be easy, especially in academic settings. Tenure and promotion for faculty members tend to privilege publications in purely academic outlets and collaborations with other academics. Fostering trusting relationships between academic researchers and community groups takes time and significant commitment. While the challenges are great, practitioners of citizen science have begun to address this issue more explicitly than before. There is growing recognition that gender, race, ethnicity, and socioeconomic backgrounds shape participation in citizen science. A leading scholar on citizen science in Europe, Muki Haklay, has written on the need for explicit and purposeful inclusion strategies for marginalized groups.[3] The U.S. Citizen Science Association established a working group on inclusion, diversity, and justice, recognizing that barriers for engagement exist.[4] Further efforts along these lines will be needed to ensure that as citizen science becomes institutionalized, it does not also become exclusionary.

Participatory environmental research can call attention to the unfair distribution of pollution burdens and the ways that social inequality and power relations shape environmental problems. In the case of fracking, grassroots air monitoring is revealing that people living in extractive communities—many of them economically marginalized and geographically isolated—are exposed to toxic pollution. Likewise, radiation monitoring in Japan is showing the extent to which certain populations are burdened with radioactive contamination. Some of the CRMOs have been working with the people who were displaced by the nuclear accident in their pursuit of legal remedies. Soil contamination data, for instance, has been used in litigation to argue for the better protection of rights of the displaced people.

Citizen science most effectively supports environmental justice when it is done in collaboration with grassroots organizations representing oppressed and marginalized groups. In the Mexican case of GE corn, activists linked contamination of traditional corn with the oppression of indigenous peoples and peasants. Grassroots organizations called for collaboration with scientists to learn more about the extent of unwanted commingling of GE and traditional corn varieties. For the previous decade, peasant organizations had been contesting the reduction of farmer-support policies and the negative impacts of trade liberalization on small-scale corn growers. Indigenous groups were simultaneously demanding recognition of their rights to autonomy and respect for their cultures. The citizen science project helped the struggle for justice, as it provided additional evidence that the multinational-dominated global food system threatened rural livelihoods and cultures. Without the central involvement of peasant and indigenous organizations, the story of GE corn in Mexico might not have been told in a way that highlighted broader justice concerns.

Beyond the examples we examined in previous chapters, there are models of participatory research that involve not only probing the biophysical aspects of a particular environmental issue but also asking questions about equity, power, and justice. For instance, Barbara L. Allen, a social scientist, collaborated on a community-based participatory environmental health study in the industrial port area of Marseille, France. For years, residents claimed they were experiencing excessive illnesses related to hundreds of gas, chemical, and steel installations in the area, yet official health data did not support these claims, made by largely working-class residents. Allen worked with both epidemiologists and residents to systematically collect, interpret, and contextualize health data. The results indicated a higher prevalence of asthma and cancer compared to the rest of the country. The process of research was "strongly participatory" at each step, leading to what Allen calls "knowledge justice." She states, "First, their observations as residents are included as the basis of the questions asked in the health survey. Second, after the random sample is collected and the initial data are generated, we worked in focus groups with the local population, including local doctors, to interpret the data in context and produce further analyses that include their voices as 'sense makers' of the numbers. Finally, the focus group participants were asked to think collectively about 'next steps'—how might this new report validating their suspicions be used to better their environment. They came up with dozens of ideas." Allen observes that the study "is already doing 'work' for the town by challenging re-permitting applications and arguing for expanded health clinics, among other actions."[5]

However, it would be a mistake to conclude from these examples that citizen science generally calls attention to the unfair distribution of pollution burdens or helps people challenge unjust procedures for making environmental decisions. Most of the forms of citizen science that are now being promoted on SciStarter or through academic and government science institutions give little attention to justice or social equity. For this reason, many environmental justice activists are rightly skeptical of the dominant conversation about citizen science that emphasizes how it can help professional scientists or conservationists.

It may be difficult to broaden commitments to environmental justice in the growing citizen science community. Citizen science started by or in close engagement with social justice organizations usually has different goals than academic, regulatory, or conservation science. Grassroots scientists typically want to produce observations that are actionable in terms of creating media attention, getting a foot in regulatory negotiations, or pursuing litigation. Working toward applied outcomes of this sort can feel scary or uncomfortable for many academic scientists, particularly when they are threatened with accusations of bias or partisanship. Scientists in government agencies also experience pressure to remain politically neutral in their assessments. Another challenge is that relationships with oppressed and marginalized groups require intentional and long-term fostering of trust and respect—a process that does not map onto the fast-paced grants cycle that characterizes much academic science. Furthermore, many communities have experienced research that was exploitative or purely extractive, only benefiting paid researchers. Building mutually beneficial and trusting relationships between professional scientists and communities seeking environmental justice will require reforms at the level of institutions, including universities, funding agencies, and other scientific organizations.

GIVING VOICE

The third virtue that we see in participatory environmental research is its capacity to give public voice to participants at a time when scientific knowledge and technical expertise are crucial to many important decisions. Although it may appear to be under attack by religious and business forces, science remains central to policy and regulatory processes in countries around the world.[6] Citizen science provides an entry point into the highly technical worlds of environmental regulation.

In Japan, for example, citizens who are monitoring GE crops have been meeting with staff from the Ministry of Environment and Ministry of Agriculture in order to share the data from their citizen science projects. This data—taken across the nation over a decade—is of interest to scientists in these agencies and functions as a launching point for citizens to relay their wider concerns about bringing GE crops to Japan. Monitoring groups were instrumental in shedding light on the possibility of gene contamination and its impacts on biodiversity from imported GE rapeseed. Because Japan does not commercially grow GE crops, the perception was that biodiversity threats from such crops were nonexistent. These groups alerted the government and scientists to the expansiveness of the problem of GE spillage and used citizen science–derived data to make their concerns about GE crops credible in the eyes of bureaucrats and experts. Every year they visit government agencies, urging them to enhance the regulations on GE crops.

Beyond granting access to regulatory decision-making, citizen science can be an opportunity for those who have experienced social injustice to amplify their voices. Even when citizen scientists are not given a seat at the table where decisions are made, participatory research can support collective actions that insist on systemic changes in policy, economic structure, and legal systems. As we saw in Mexico, research by rural community organizations revealed some of the effects of globalization in the food system. Citizen science was by no means the only action taken by the Network in Defense of Maize, but it helped build partnerships, mobilize rural communities, and challenge authorities.

Strengthening political participation through citizen science doesn't have to mean contentious social movements (though that is certainly one possibility); it can also mean learning from one another, clarifying values, and speaking publicly. A widely cited virtue of citizen science is that it can improve scientific literacy and help people develop a greater appreciation for how knowledge of the natural world is produced. Ostensibly, greater participation in citizen science could lead to more well-reasoned discussions and decisions about the crucial environmental issues of our time.

Despite the many positive ways that environmental citizen science can strengthen political participation, there is still a widely held view that science should remain "untainted" by politics. As we have noted, this creates dilemmas for citizen scientists. For example, oil and gas industry advocates have sought to discredit those who engage in environmental studies that suggest that fracking is harmful. Doubts about the credibility of a study can diminish its impacts on public policy and regulatory decision-making. Citizen scientists can take steps to legitimize their research by collaborating with credentialed scientists and laboratories, using monitoring methods that are commonly used by regulators and scientists, triangulating their data with other sources, and being transparent about their processes.

However, the pursuit of credibility may end up silencing citizen scientists rather than giving them voice if science is perceived to be incompatible with politics. For instance, Safecast, a radiation monitoring network in post–nuclear disaster Japan, has clearly staked out a position of political neutrality. Its website states, "Safecast is neither anti-nuclear nor pro-nuclear—we are pro-data. Data is apolitical."[7] This is in contrast with other measuring organizations that are gathering data with explicit objectives—for instance, to help people in contaminated regions secure their right to evacuate and to press the Japanese government for more decontamination efforts.

In some contexts, citizen scientists seek to avoid contentious politics and want to voice their concerns in ways that are more socially acceptable. For example, some people who volunteered to monitor streams for the impacts of fracking did so because it seemed to be a way to protect the environment without overtly taking a side. Small towns in Pennsylvania tend to be politically conservative, and many individuals are profiting from the natural gas industry, so there are social barriers to openly opposing the industry or calling for environmental protections. Furthermore, rural communities often lack civil society organizations such as unions, environmental advocacy groups, or cultural associations that could provide support to those who have a critical stance on the industry. In this context, volunteers in water monitoring projects sometimes commented that they were not comfortable with activism but thought they could make a difference through science.

In our view, science itself is contentious, or pregnant with the possibility of controversy. Research questions are shaped by disciplinary history and social contexts, and the interpretation of research results for environmental governance is constrained by regulatory conventions and political conditions. Human decisions and values affect all stages of the research process. By increasing the transparency of these choices, participatory science has the potential to help people take a position on decisions that matter.

Going Forward

As citizen science is rapidly being institutionalized, what will become of the transformative promise we have outlined here? In these final paragraphs, we offer some suggestions for scientists, university administrators, funding bodies, nonprofit organizations, activists, volunteers, journalists, and others who are invested in the future of participatory research.

First, let's consider how science advocates can best advance a transformative vision of citizen science. The arrival of the Trump administration, with its attacks on climate science, reinforced the idea that antiscientific ideas are on the rise. In this context, a call to "depoliticize science" has gained strong momentum. For instance, the March for Science in April 2017 reportedly attracted millions of people in more than six hundred cities around the world. The central tenet of the march was to advocate for science-based policy and to oppose the defunding of scientific research. This idea mobilized ordinary people by the thousands in support of scientific research and thinking.

In an effort to counter the Trump administration's disregard for environmental science, science marchers sometimes projected an ideal of science as separate from values and social priorities. While it might seem that the opposite of corrupt science is value-free science, the "value-free ideal" is both philosophically unsound and harmful to those who seek to use science for environmental and social justice.[8] Successful examples of environmental citizen science demonstrate the merits of science that is transparently guided by a commitment to sustainability and social change. Those of us who oppose harmful environmental policies must speak against them both with scientific reasoning and with our ethical convictions. As advocates for environmental citizen science, it is important to resist simplistic arguments against "politicizing" science and instead ask "Who is this science for?" and "What value judgments does it entail?"[9]

We hope that political leaders will draw lessons from this book. Because citizen science is an increasingly popular concept, there may be a temptation to jump into creating and funding citizen science projects. A word of caution here, though. Policy makers need to consider whether the citizen science funding is creating an undesirable trade-off with universal, publicly funded environmental and health monitoring and scientific research. We have shown that participatory projects may be distributed in ways that leave problematic gaps. For example, volunteer water monitoring projects are not present in every watershed, and voluntary radiation monitoring efforts are limited to locations where people can commit time and resources to data collection. These gaps can lead to faulty conclusions about the prevalence of environmental problems as well as social inequities when local communities lack scientific data about their environmental exposures. The question for policy makers must be, Is your citizen science project supplementing—not replacing or shrinking—regularized, long-term, and publicly resourced programs for protecting the environment and well-being of communities?

Projects that regard volunteers merely as data collectors rather than collaborators who bring necessary knowledge and perspectives should be a lower priority for public dollars. Instead, funding could be directed toward participant-led and truly collaborative projects, steering away from the deficit model of public understanding of science. Citizen science is, certainly, an excellent means of science education, but participants can also educate project leaders and each other about their own experience-based knowledge. In many cases, communities are already mobilizing and collectively acting to identify and solve environmental problems. Public policy can support community-based citizen science in multiple ways. For example, science funding programs could encourage universities and environmental agencies to hire liaisons from communities who struggle with environmental injustices.

Political leaders and regulators also need to consider how to use knowledge produced through citizen science when they make decisions. Regulatory processes could be structured to take into account the environmental knowledge that is produced through collaborative and participant-led research. Rather than discounting this knowledge as lacking credibility, recognize that constituents who are committed to observing their environments bring invaluable perspectives and understandings of environmental policy issues to the discussion. Indeed, they may know things that professional scientists do not by virtue of their intimate familiarity with particular locales and situations. Furthermore, their interpretation of data may include systemic, historic, and institutional issues that create environmental burdens. Such issues are often left out of regulatory processes to the detriment of marginalized communities.

Beyond public policy, academic research scientists have a crucial role to play in shaping the future of citizen science. Incentives are growing for research scientists to incorporate citizen scientists into their projects. We encourage scientists to give more thought not only to data quality but also to the issues of voice, power, and diversity. Even when the project is ostensibly about a limited set of data that may not seem to involve any of these issues on the surface, there may be an opportunity to contextualize the study politically, culturally, and historically.

Within universities, interdisciplinary collaboration is a means to develop more robust and transformative citizen science initiatives. Professional scientists may feel that focusing on issues of justice, power, history, scale, and context of environmental problems gets in the way of doing good science. It may feel overwhelming to have to consider such a wide range of issues in addition to trying to ensure scientific validity and reliability. This is one good reason to find interdisciplinary research partners. Social scientists are likely to be more capable of providing analyses of the social, political, and historical context for the environmental issue at the center of your study—as in the example of Barbara L. Allen's work with epidemiologists and French working-class communities, described previously. Furthermore, partners in civil society, such as policy advocacy organizations, may also have the expertise needed to identify the social groups who should be included in a participatory study, the environmental justice issues at stake, and the opportunities to connect citizen science to political debates.

Scientists may also think about expanding the goals of the project to include developing participants' political voices to challenge systemic environmental injustice. For example, an environmental monitoring project might reveal a problem requiring policy change or improved regulatory enforcement. In cases like this, volunteers are not just data collectors but also politically savvy agents who may be able to carry out sustained advocacy. Even though professional scientists may not wish to lead an advocacy campaign themselves, they can go beyond the usual forms of education associated with citizen science projects by disseminating information to volunteers about other people, campaigns, and organizations that work on related issues.

As we suggested earlier, there is also important work to be done by research administrators and funders, who are in the position to shift the priorities and practices of citizen science. Currently, citizen science projects based at universities rarely involve participants at the agenda-setting and grant-writing stage. Projects are often guided by academic values—getting grants and publishing articles. Could universities become more inclusive sites where typically marginalized groups are able to pose questions and call for research on problems that they identify?

One example shows how it can be done. A sociologist and an entomologist at the University of Wisconsin organized a series of deliberations that included people who had diverse perspectives on the problem of honeybee colony collapse—a controversial issue in agricultural communities because of suspicions that agricultural pesticides are harming the bees. Together they identified research questions and developed experimental study designs that reflected the knowledge and ideas that participants (beekeepers, growers, scientists, and regulators) brought to the project. Some participants, such as beekeepers, helped collect data during the experiment. Later, further deliberations about the results led to a new study design. While this participatory research process was not fully open to all members of the public, it did successfully bring the perspectives of nonscientists (beekeepers and farmers) into the earliest stages of designing an experimental study about the threats to honeybees. Furthermore, the deliberative process reduced tensions among the stakeholder groups and generated mutual respect for the knowledge that each stakeholder had to offer.[10]

While this example suggests a model for collaboration, communities' interests will not always align with the research questions that interest professional scientists and lead to peer-reviewed publications. Yet incentives and rewards for academic research are shifting. It is useful to remember that many universities now recognize the commercialization of science through patenting and licensing as an important faculty contribution—a significant shift in what counts as productivity. Citizen science might not reward universities with financial gains, but in the same way that patenting came to be valued in academia, universities could also reconsider the prevailing perception that community-based collaboration is only an auxiliary to the "real" research. Indeed, civic engagement is increasingly recognized as a value that universities provide to broader society, and here we see an important opportunity to reward scientific work that makes space for excluded voices and perspectives.

It remains to be seen how universities might successfully promote citizen science that is transformative, and it will likely vary by institution. On one hand, public universities have a clear mandate to serve the public; on the other hand, some private universities are less dependent on external funding, perhaps making them more flexible in experimenting with new ways of doing citizen science.

Academia is just one site where environmental citizen science is formulated. What about environmental activist groups and nonprofits that initiate citizen science projects? Our case studies highlighted dilemmas that can make participatory research a double-edged sword for organizations working for social change. What can activists and those in the nonprofit sector do to help advance more transformative approaches to citizen science? One important task is to contextualize scientific data. Many citizen science projects rationalize environmental issues, reducing complex social problems to scientific measurements, but activists can use storytelling, photography, argumentation, and many other methods of situating the observations gathered through citizen science in their social and political contexts. Contextualizing data should include attention to how the scale of data collection aligns with the different scales of the problem. In particular, it is worthwhile to consider how data that are at the level of individuals, households, and neighborhoods might be "scaled up" to make arguments for systemic changes in environmental law and policy. We have seen that the research process can be a solidarity-building method for social movements. Sharing citizen science projects with other groups can spotlight systemic environmental justice issues that span locales.

What if you are reading this as a citizen scientist—or potential volunteer—yourself? As we have stressed in the book, it is important to consider the values that are woven into the fabric of projects, even if they ostensibly have no political agenda. Recall the project by Biology Fortified Inc. in chapter 5. On the surface, this project appeared neutral, but the organization's advocacy for agricultural biotechnology was implicit in the research objectives and design of the study. This is not to say that citizen science projects that have a political objective are "biased" and hence not scientific. Our point is that citizen scientists should carefully consider how their values align with those of the sponsoring group. Every citizen science project has its own value framing, regardless of whether it is explicit or implicit in their description of the project. Before signing up, volunteers should investigate what objectives and assumptions are embedded in the project. Volunteers can also talk to citizen science organizers about how a project could be expanded or changed to better suit their own goals.

In addition, individual citizen scientists can insist on greater inclusion and diversity of volunteers. Are volunteers mostly all one gender, racial or ethnic group, or social class? Is this by design (e.g., a study meant to be by and for a particular marginalized group) or is it because of barriers to wider participation? Citizen scientists can make crucial contributions by questioning project leaders about these exclusions.

We also see a potential role for the media in shifting public understanding of what environmental citizen science is and does. Science journalists frequently write about citizen science projects without investigating the meaning of participation by volunteers. When reporting on citizen science projects that are initiated by scientists, reporters can probe more deeply about the degree of "participation" by volunteers. Journalistic reports can more critically examine the degree to which diverse participants are recruited into the project, whether volunteers are allowed to set the research agenda and direct data interpretation, and how participation links to political action or other outcomes that are important for society. Journalists can also add contextual information for citizen science projects, such as funding constraints that necessitate volunteers in the first place or the areas of research that struggle to attract willing volunteers. Finally, journalists who write about politics, and not only science writers, should pay attention to citizen science—particularly when activists use participatory research to challenge dominant narratives about environmental policy and regulations.

We offer these suggestions while recognizing wide variation in the settings in which citizen science takes place. Skillfully navigating the dilemmas that inevitably arise requires more than a few context-specific tips and strategies; it takes a sociological imagination. Few of the people we encountered in the research for this book had any training in sociology, yet savvy citizen scientists were able to think through the connections between the troubles they were experiencing and broader public issues to figure out the best way to proceed. Their insights and actions leave us with optimism about the prospects for more transformative practices of participatory research. We do not naively believe that citizen science is making democracy stronger or revolutionizing science. Yet we do see real value in taking science off of its pedestal and regarding it as something we all can do to make sense of the world in which we live.

Appendix

Resources for Getting Involved in Citizen Science

Professional Associations

Citizen science is increasingly professionalized. There are now professional associations in many parts of the world, which hold annual or biennial meetings and host other activities such as online discussions and workshops. Some of these also have a presence on Facebook and Twitter.

Australian Citizen Science Association
https://citizenscience.org.au

CitizenScience.Asia
https://www.facebook.com/CitSciAsia

United States' Citizen Science Association
http://www.citizenscience.org

European Citizen Science Association
https://ecsa.citizen-science.net

Nonprofit Organizations

While this list is far from exhaustive, these are some of the non-profit organizations that have developed and shared novel participatory research tools and methods of community engagement.

All India People's Science Network
http://aipsn.net

Citizens' Nuclear Information Center, Japan
http://www.cnic.jp

Communities for a Better Environment
http://www.cbecal.org

Fundación Ciencia Ciudadana (Citizen Science Foundation), Chile
http://cienciaciudadana.cl

Louisiana Bucket Brigade
http://www.labucketbrigade.org

Minnna No Data Site, Japan
https://minnanods.net

National Environmental Education Foundation (NEEF)
https://www.neefusa.org/citizen-science

No! GMO Campaign, Japan
http://gmo-iranai.lolipop.jp/index.php

Pesticide Action Network
http://www.panna.org

Public Lab
https://publiclab.org

Safecast
https://blog.safecast.org

Programs Based in Universities and Research Centers

Environmental citizen science takes place under the leadership of scientists and staff at many different colleges, universities, and research centers. These are just some of the centers that have had a sustained commitment to citizen science. The websites for many of these programs provide useful resources, such as guides to best practices, links to projects seeking citizen scientists, and so on.

Alliance for Aquatic Resource Monitoring (ALLARM), Dickinson College
https://www.dickinson.edu/allarm

Center for Community and Citizen Science, University of California Davis
https://education.ucdavis.edu/community-and-citizen-science

Citizen Sense at Goldsmiths, University of London
https://citizensense.net

CitSci, Colorado State University
http://www.citsci.org

Cornell Lab of Ornithology
http://www.birds.cornell.edu

Extreme Citizen Science (ExCiteS), University College London
http://www.ucl.ac.uk/excites

Fair Tech Collective, Drexel University
https://www.fairtechcollective.org

Nouminren (Japan Family Farmers' Movement) Food Research
    Laboratory
http://earlybirds.ddo.jp/bunseki/index.html

Red Chilena de Ciencia Ciudadana (Chilean Citizen Science
    Network), Centro de Estudios Avanzados en Zonas Áridas
    (CEAZA; Center for Advanced Studies in Arid Zones)
http://cienciaciudadana.ceaza.cl/red-chilena-ciencia-ciudadana.html

SciStarter, University of Arizona
https://scistarter.com

Social Science Environmental Health Research Institute, Northeast-
    ern University
https://www.northeastern.edu/environmentalhealth

Government Resources

In the United States, federal agencies now encourage citizen science. Here are some of the websites where people can be connected
to citizen science projects that are relevant to environmental monitoring and regulation.

Official U.S. government website
https://www.citizenscience.gov

U.S. Environmental Protection Agency
https://www.epa.gov/citizen-science

U.S. Forest Service
https://www.fs.fed.us/working-with-us/citizen-science

Other countries also have various citizen science programs affiliated with government agencies.

Vigie-Nature (with the Natural History Museum), France
http://vigienature.mnhn.fr/blog

Natural History Museum, United Kingdom
http://www.nhm.ac.uk/take-part/citizen-science.html

Environmental Observation Framework, United Kingdom
http://www.ukeof.org.uk/our-work/citizen-science

National Parks Board, Singapore
https://www.nparks.gov.sg/biodiversity/community-in-nature
 -initiative/citizen-science-programmes

Academic Journals

Research using citizen science is published in a wide array of journals. Here, we highlight two journals that make citizen science research accessible to readers online.

Citizen Science: Theory and Practice
https://theoryandpractice.citizenscienceassociation.org
An open-access, peer-reviewed journal published on behalf of the
 Citizen Science Association

Citizen Science Today
http://www.citizensciencetoday.org
An online publication that "highlights good writing and clear-
 sighted scholarship related to citizen science"

Television Series

This PBS series showcases a wide array of examples of citizen science, including some of the cases considered in this book. The episodes can be viewed online, and there are also links to projects you can join.

"Crowd and Cloud"
http://crowdandcloud.org/about-the-series

Notes

1. Environmental Citizen Science

1. Sebastián Ureta, "Baselining Pollution: Producing 'Natural Soil' for an Environmental Risk Assessment Exercise in Chile," *Journal of Environmental Policy & Planning* 7200 (November 2017): 1–14, https://doi.org/10.1080/1523908X.2017.1410430.

2. Deborah J. Trumbull et al., "Thinking Scientifically during Participation in a Citizen-Science Project," *Science Education* 84 (2000): 265–275, https://doi.org/10.1002/(SICI)1098-237X(200003)84:2<265::AID-SCE7>3.3.CO;2-X.

3. For example, in 2015, the National Parks Board of Singapore launched Community in Nature Citizen Science programs to involve the community in biodiversity monitoring. Projects include the Marine Eco-toxicity Biomonitoring Programme and Butterfly Watch.

4. Philip Mirowski, "Against Citizen Science," *Aeon*, no. 20 (November 2017), https://aeon.co/essays/is-grassroots-citizen-science-a-front-for-big-business.

5. Max Craglia and Lea Shanley, "Data Democracy—Increased Supply of Geospatial Information and Expanded Participatory Processes in the Production of Data," *International Journal of Digital Earth* 8, no. 9 (September 2015): 679–693, https://doi.org/10.1080/17538947.2015.1008214.

6. Jenn Gustetic, Kristen Honey, and Lea Shanley, "Open Science and Innovation: Of the People, by the People, for the People," Obama White House, 2015, https://obamawhitehouse.archives.gov/blog/2015/09/09/open-science-and-innovation-people-people-people.

7. "The Open Government Partnership: Second Open Government National Action Plan for the United States of America: A Preview Report," Obama White House, October 31, 2013, https://obamawhitehouse.archives.gov/sites/default/files/docs/preview_report_of_open_gov_national_action_plan_10_30.pdf.

8. The project website can be viewed at http://www.socientize.eu. Fermín Serrano Sanz et al., *White Paper on Citizen Science for Europe* (Brussels: Commission européenne, 2014).

9. Muki Haklay, *Citizen Science and Policy: A European Perspective*, vol. 4 (Washington, D.C.: Woodrow Wilson Center, 2015).

10. Janis L. Dickinson and Rick Bonney, *Citizen Science: Public Participation in Environmental Research* (Ithaca, N.Y.: Cornell University Press, 2012); Rick Bonney et al., "Citizen Science: Next Steps for Citizen Science," *Science* 343, no. 6178 (2014): 1436–1437, https://doi.org/10.1126/science.1251554.

11. Rick Bonney, Jennifer Shirk, and Tina B. Philip, "Citizen Science," in *Encyclopedia of Science Education*, ed. Richard Gunstone (Dordrecht: Springer, 2015), 152–154.

12. Kathleen Toerpe, "The Rise of Citizen Science," *Futurist* 47, no. 4 (2013): 25–30.

13. Alan Irwin, *Citizen Science: A Study of People, Expertise and Sustainable Development* (London: Routledge, 1995).

14. Barbara L. Allen, *Uneasy Alchemy: Citizens and Experts in Louisiana's Chemical Corridor Disputes* (Cambridge, Mass.: MIT Press, 2003); Gwen Ottinger, *Refining Expertise: How Responsible Engineers Subvert Environmental Justice Challenges* (New York: New York University Press, 2013).

15. Jeremy Vetter, "Introduction: Lay Participation in the History of Scientific Observation," *Science in Context* 24, no. 2 (April 28, 2011): 127–141, https://doi.org/10.1017/S0269889711000032.

16. "What Is the COOP Program? NOAA's National Weather Service," NWS Cooperative Observer Program, 2014, http://www.nws.noaa.gov/om/coop/what-is-coop.html.

17. Vincent Devictor, Robert J. Whittaker, and Coralie Beltrame, "Beyond Scarcity: Citizen Science Programmes as Useful Tools for Conservation Biogeography," *Diversity and Distributions* 16, no. 3 (2010): 354–362; Mark D. Schwartz, Julio L. Betancourt, and Jake F. Weltzin, "From

Caprio's Lilacs to the USA National Phenology Network," *Frontiers in Ecology and the Environment* 10, no. 6 (2012): 324–327.

18. Aya H. Kimura, "Citizen Monitoring in Japan: Spiderwort and Cherry-blossoms," article in preparation, n.d.

19. Max Pfeffer and Linda Plummer Wagenet, "Volunteer Environmental Monitoring, Knowledge Creation and Citizen–Scientist Interaction," in *Sage Handbook on Environment and Society*, 2007, 235–249.

20. Vickie Curtis, "Motivation to Participate in an Online Citizen Science Game: A Study of Foldit," *Science Communication* 37, no. 6 (2015): 723–746, https://doi.org/10.1177/1075547015609322.

21. Barbara L. Allen, "Strongly Participatory Science and Knowledge Justice in an Environmentally Contested Region," *Science, Technology & Human Values* 43, no. 6 (2018): 949, https://doi.org/10.1177/0162243918758380.

22. David Bidwell, "Is Community-Based Participatory Research Post-normal Science?," *Science, Technology & Human Values* 34, no. 6 (2009): 741–761, https://doi.org/10.1177/0162243909340262.

23. Phil Brown, "When the Public Knows Better: Popular Epidemiology," *Environment* 35, no. 8 (1993): 16–21.

24. Sarah Elwood, "Negotiating Knowledge Production: The Everyday Inclusions, Exclusions, and Contradictions of Participatory GIS Research," *Professional Geographer* 58, no. 2 (2006): 197–208, https://doi.org/10.1111/j.1467-9272.2006.00526.x; Joe Bryan, "Walking the Line: Participatory Mapping, Indigenous Rights, and Neoliberalism," *Geoforum* 42, no. 1 (2011): 40–50, https://doi.org/10.1016/j.geoforum.2010.09.001.

25. Matt Ratto, "Critical Making: Conceptual and Material Studies in Technology and Social Life," *Information Society* 27, no. 4 (2011): 252–260; Sara Ann Wylie et al., "Institutions for Civic Technoscience: How Critical Making Is Transforming Environmental Research," *Information Society* 30, no. 2 (2014): 116–126, https://doi.org/10.1080/01972243.2014.875783.

26. Anna Lawrence, "'No Personal Motive?' Volunteers, Biodiversity, and the False Dichotomies of Participation," *Ethics, Place & Environment* 9, no. 3 (2006): 279–298, http://www.tandfonline.com/doi/abs/10.1080/13668790600893319; Cristina Gouveia and Alexandra Fonseca, "New Approaches to Environmental Monitoring: The Use of ICT to Explore Volunteered Geographic Information," *GeoJournal* 72, no. 3–4 (2008): 185–197, https://doi.org/10.1007/s10708-008-9183-3; Bidwell,

"Community-Based Participatory Research"; Jennifer Shirk et al., "Public Participation in Scientific Research: A Framework for Deliberate Design," *Ecology and Society* 17, no. 2 (2012): 29, https://doi.org/10.5751/ES-04705-170229; Muki Haklay, "Citizen Science and Volunteered Geographic Information: Overview and Typology of Participation," in *Crowdsourcing Geographic Knowledge: Volunteered Geographic Information (VGI) in Theory and Practice*, ed. Daniel Sui, Sarah Elwood, and Michael Goodchild (Dordrecht: Springer Netherlands, 2013), 105–122.

27. Andrea Wiggins and Kevin Crowston, "From Conservation to Crowdsourcing: A Typology of Citizen Science," *Proceedings of the Annual Hawaii International Conference on System Sciences* (2011): 1–10, https://doi.org/10.1109/HICSS.2011.207.

28. Website is at https://www.zooniverse.org/projects/povich/milky-way-project.

29. Barbara Prainsack and Hauke Riesch, "Interdisciplinarity Reloaded: Drawing Lessons from 'Citizen Science,'" in *Investigating Interdisciplinary Collaboration: Theory and Practice*, ed. Scott Frickel, Albert Mathieu, and Barbara Prainsack (New Brunswick, N.J.: Rutgers University Press, 2016), 151–164.

30. Ross K. Meentemeyer et al., "Citizen Science Helps Predict Risk of Emerging Infectious Disease," *Frontiers in Ecology and the Environment* 13, no. 4 (2015): 189, https://doi.org/10.1890/140299.

31. Mark Chandler et al., "Contributions to Publications and Management Plans from 7 Years of Citizen Science: Use of a Novel Evaluation Tool on Earthwatch-Supported Projects," *Biological Conservation* 208 (2017): 163–173.

32. E. J. Theobald et al., "Global Change and Local Solutions: Tapping the Unrealized Potential of Citizen Science for Biodiversity Research," *Biological Conservation* 181 (2015), https://doi.org/10.1016/j.biocon.2014.10.021.

33. Christopher Kullenberg and Dick Kasperowski, "What Is Citizen Science? A Scientometric Meta-analysis," *PLOS ONE* 11, no. 1 (2016): e0147152, https://doi.org/10.1371/journal.pone.0147152; Theobald et al., "Global Change and Local Solutions."

34. Theobald et al., "Global Change and Local Solutions."

35. H. K. Burgess et al., "The Science of Citizen Science: Exploring Barriers to Use as a Primary Research Tool," *Biological Conservation* 208 (2017): 113–120.

36. Burgess et al.

37. Theobald et al., "Global Change and Local Solutions"; Burgess et al., "Science of Citizen Science."

38. Janis L. Dickinson, Benjamin Zuckerberg, and David N. Bonter, "Citizen Science as an Ecological Research Tool: Challenges and Benefits," *Annual Review of Ecology, Evolution, and Systematics* 41, no. 1 (December 2010): 149–172, https://doi.org/10.1146/annurev-ecolsys-102209-144636; Alexandra Swanson et al., "A Generalized Approach for Producing, Quantifying, and Validating Citizen Science Data from Wildlife Images," *Conservation Biology* 30, no. 3 (June 2016): 520–531, https://doi.org/10.1111/cobi.12695.

39. Rebecca Lave, "Neoliberalism and the Production of Environmental Knowledge," *Environment and Society: Advances in Research* 3, no. 1 (December 18, 2012): 28, https://doi.org/10.3167/ares.2012.030103.

40. Mirowski, "Against Citizen Science."

41. National Research Council, *Learning Science in Informal Environments: People, Places, and Pursuits* (Washington, D.C.: National Academies Press, 2009).

42. In 2007, NSF helped fund the Center for the Advancement of Informal Science Education (CAISE), which then established the citizen science inquiry group. Its report in 2009 defined and disseminated the idea of PPSR and suggested citizen science and crowd sourcing as types of PPSR. Shirk et al., "Public Participation in Scientific Research."

43. Kristine F. Stepenuck and Linda T. Green, "Individual- and Community-Level Impacts of Volunteer Environmental Monitoring: A Synthesis of Peer-Reviewed Literature," *Ecology and Society* 20, no. 3 (2015): 19, https://doi.org/10.5751/ES-07329-200319.

44. Christine Overdevest, Cailin Huyck Orr, and Kristine Stepenuck, "Volunteer Stream Monitoring and Local Participation in Natural Resource Issues," *Research in Human Ecology* 11, no. 2 (2004): 177–185.

45. Richard Stedman et al., "Cleaning Up Water? Or Building Rural Community? Community Watershed Organizations in Pennsylvania," *Rural Sociology* 74, no. 2 (October 22, 2009): 178–200, https://doi.org/10.1111/j.1549-0831.2009.tb00388.x.

46. Maria E. Fernandez-Gimenez, Heidi L. Ballard, and Victoria E. Sturtevant, "Adaptive Management and Social Learning in Collaborative and Community-Based Monitoring: A Study of Five

Community-Based Forestry Organizations in the Western USA,"
Ecology and Society 13, no. 2 (2008): 4, http://www.mtnforum.org/sites/
default/files/pub/4143.pdf.

47. Stedman et al., "Cleaning Up Water?"

48. Overdevest, Orr, and Stepenuck, "Volunteer Stream Monitoring," 183.

49. Behnam Taebi et al., "Responsible Innovation as an Endorsement of
Public Values: The Need for Interdisciplinary Research," *Journal of
Responsible Innovation* 1, no. 1 (2014): 118–124.

50. Franziska Sattler and Claudia Göbel, "Summary of Results from Our
First #CitSciChat on CS and Responsible Research and Innovation |
European Citizen Science Association (ECSA)," n.d., https://ecsa
.citizen-science.net/blog/summary-results-our-first-citscichat-cs-and
-responsible-research-and-innovation; Arie Rip, "The Past and Future
of RRI," *Life Sciences, Society and Policy* 10, no. 1 (December 2014): 17,
https://doi.org/10.1186/s40504-014-0017-4.

51. Justin Dillon, Robert B. Stevenson, and Arjen E. J. Wals, "Introduction
to the Special Section Moving from Citizen to Civic Science to Address
Wicked Conservation Problems. Corrected by Erratum 12844," *Conservation Biology* 30, no. 3 (June 2016): 451, https://doi.org/10.1111/cobi.12689.

52. Sandra Harding, *Objectivity and Diversity: Another Logic of Scientific
Research* (Chicago: University of Chicago Press, 2015).

53. Daniela Soleri et al., "Finding Pathways to More Equitable and Meaningful Public-Scientist Partnerships," *Citizen Science: Theory and Practice*
1, no. 1 (2016): 1–11, http://doi.org/10.5334/cstp.46.

54. The project website is at http://listening.coweeta.uga.edu/about;
Brian J. Burke and Nik Heynen, "Transforming Participatory Science
into Socioecological Praxis: Valuing Marginalized Environmental
Knowledges in the Face of Neoliberalization of Nature and Science,"
Environment and Society: Advances in Research 5 (2014): 8, https://doi.org/
10.3167/ares.2014.050102.

55. Gwen Ottinger, "Social Movement-Based Citizen Science," in *The
Rightful Place of Science: Citizen Science*, ed. Darlene Cavalier and Eric B.
Kennedy (Tempe, Ariz.: Consortium for Science, Policy, and Outcomes,
2016), 89–103; Phil Brown, *Toxic Exposures: Contested Illnesses and the
Environmental Health Movement* (New York: Columbia University Press,
2007).

56. Sabrina McCormick, "Transforming Oil Activism: From Legal Constraints to Evidenciary Opportunity," *Sociology of Crime Law and Deviance* 17 (2012): 113–131.

57. Wylie et al., "Institutions for Civic Technoscience"; Jessica Breen et al., "Mapping Grassroots: Geodata and the Structure of Community-Led Open Environmental Science," *ACME: An International E-journal for Critical Geographies* 14, no. 3 (2015): 849–873.

58. Jason Corburn, *Street Science: Community Knowledge and Environmental Health Justice* (Cambridge, Mass.: MIT Press, 2005), 94–109.

59. Dara O'Rourke and Gregg P. Macey, "Community Environmental Policing: Assessing New Strategies of Public Participation in Environmental Regulation," *Journal of Policy Analysis and Management* 22, no. 3 (2003): 383–414, https://doi.org/10.1002/pam.10138; Gwen Ottinger, "Buckets of Resistance: Standards and the Effectiveness of Citizen Science," *Science, Technology & Human Values* 35, no. 2 (June 12, 2010): 244–270, https://doi.org/10.1177/0162243909337121.

60. Jennifer Gabrys, Helen Pritchard, and Benjamin Barratt, "Just Good Enough Data: Figuring Data Citizenships through Air Pollution Sensing and Data Stories," *Big Data & Society* 3, no. 2 (2016), https://doi.org/10.1177/2053951716679677.

61. Julia Frost Nerbonne and Kristen C. Nelson, "Volunteer Macroinvertebrate Monitoring in the United States: Resource Mobilization and Comparative State Structures," *Society & Natural Resources* 17, no. 3 (September 2004): 817–839, https://doi.org/10.1080/08941920490493837; Julia Frost Nerbonne et al., "Effect of Sampling Protocol and Volunteer Bias When Sampling for Macroinvertebrates," *Journal of the North American Benthological Society* 27, no. 3 (2008): 640–646.

62. Abby Kinchy, Kirk Jalbert, and Jessica Lyons, "What Is Volunteer Water Monitoring Good For? Fracking and the Plural Logics of Participatory Science," *Political Power and Social Theory* 27 (2014): 259–289, https://doi.org/10.1108/S0198-871920140000027017.

63. Chandler et al., "Contributions to Publications."

64. The political sociology of science examines the dynamics, patterns of interactions, and sustained powers of scientific institutions. Our methodology closely aligns with this approach; we are interested in understanding the relationships and interactions between citizen science

and social institutions (such as political, economic, educational, and civic institutions). See, for example, Kelly Moore et al., "Science and Neoliberal Globalization: A Political Sociological Approach," *Theory and Society* 40, no. 5 (2011): 505–532; Scott Frickel and Kelly Moore, eds., *The New Political Sociology of Science: Institutions, Networks, and Power* (Madison: University of Wisconsin Press, 2005).

65. C. Wright Mills, *The Sociological Imagination*, 40th anniversary ed. (New York: Oxford University Press, 2000), 5.

2. How Is Environmental Citizen Science Political?

1. Mara Gittleman, Kelli Jordan, and Eric Brelsford, "Using Citizen Science to Quantify Community Garden Crop Yields," *Cities and the Environment (CATE)* 5, no. 1 (2012): 1–14.

2. See the Farming Concrete website, https://farmingconcrete.org/toolkit/.

3. Melissa Leach, Ian Scoones, and Brian Wynne, *Science and Citizens: Globalization and the Challenge of Engagement* (London: Zed Books, 2005).

4. Iris Marion Young, "Polity and Group Difference: A Critique of the Ideal of Universal Citizenship," *Ethics* 99 (1989): 250–274. See also Iris Marion Young, *Inclusion and Democracy* (Cambridge: Oxford University Press, 2000).

5. E. F. Isin and P. Nyers, *Routledge Handbook of Global Citizenship Studies* (London: Routledge, 2014), 1.

6. Sandra Harding, *Whose Science? Whose Knowledge?* (Ithaca, N.Y.: Cornell University Press, 1991); Sandra Harding, *Science and Social Inequality: Feminist and Postcolonial Issues* (Champaign: University of Illinois Press, 2006).

7. Scott Frickel, Richard Campanella, and M. Bess Vincent, "Mapping Knowledge Investments in the Aftermath of Hurricane Katrina: A New Approach for Assessing Regulatory Agency Responses to Environmental Disaster," *Environmental Science & Policy* 12, no. 2 (2009): 119.

8. Jason Owen-Smith, "Commercial Imbroglios: Proprietary Science and the Contemporary University," in *The New Political Sociology of Science: Institutions, Networks, and Power*, ed. Scott Frickel and Kelly Moore (Madison: University of Wisconsin Press, 2006), 63–90.

9. David J. Hess, *Alternative Pathways in Science and Technology: Activism, Innovation, and the Environment in an Era of Globalization* (Cambridge,

Mass.: MIT Press, 2007); Scott Frickel et al., "Undone Science: Charting Social Movement and Civil Society Challenges to Research Agenda Setting," *Science, Technology & Human Values* 35, no. 4 (October 2009): 444–473, https://doi.org/10.1177/0162243909345836.

10. Gwen Ottinger, "Buckets of Resistance: Standards and the Effectiveness of Citizen Science," *Science, Technology & Human Values* 35, no. 2 (June 12, 2010): 244–270, https://doi.org/10.1177/0162243909337121.

11. There is a rich literature that critiques homogenizing views of "local communities" in projects promoting community-based resource management and local knowledge. See Carl Wilmsen et al., eds., *Partnerships for Empowerment: Participatory Research for Community-Based Natural Resource Management* (London: Earthscan, 2008), https://doi.org/10.4324/9781849772143; Arun Agrawal and Clark C. Gibson, *Communities and the Environment: Ethnicity, Gender, and the State in Community-Based Conservation* (New Brunswick, N.J.: Rutgers University Press, 2001).

12. Heidi L. Ballard and Brinda Sarathy, "Inclusion and Exclusion: Immigrant Forest Workers and Participation in Natural Resource Management," in *Partnerships for Empowerment*, ed. Wilmsen et al., 179.

13. Jill Lindsey Harrison, "Parsing 'Participation' in Action Research: Navigating the Challenges of Lay Involvement in Technically Complex Participatory Science Projects," *Society & Natural Resources* 24, no. 7 (2011): 702–716, https://doi.org/10.1080/08941920903403115.

14. Robert D. Bullard, *Confronting Environmental Racism: Voices from the Grassroots* (Boston: South End Press, 1983); Robert W. Williams, "Environmental Injustice in America and Its Politics of Scale," *Political Geography* 18, no. 1 (January 1999): 49–73, http://dx.doi.org/10.1016/S0962-6298(98)00076-6.

15. Greg Newman et al., "User-Friendly Web Mapping: Lessons from a Citizen Science Website," *International Journal of Geographical Information Science* 24, no. 12 (November 26, 2010): 1851–1869, https://doi.org/10.1080/13658816.2010.490532.

16. Nina Eliasoph, *Making Volunteers: Civic Life after Welfare's End* (Princeton: Princeton University Press, 2011), x.

17. Eliasoph, 56.

18. Andrew Szasz, "Progress through Mischief: The Social Movement Alternative to Secondary Associations," *Politics & Society* 20, no. 4 (1992): 521–528; Nina Eliasoph and Paul Lichterman, "Making Things Political," in *Handbook of Cultural Sociology*, ed. John R. Hall, Laura Grindstaff, and Ming-Cheng Lo (London: Routledge, 2010), 483–493.

19. See the EPA's volunteer monitoring program, https://archive.epa.gov/water/archive/web/html/epasvmp.html.

20. Max Pfeffer and Linda Plummer Wagenet, "Volunteer Environmental Monitoring, Knowledge Creation and Citizen–Scientist Interaction," in *Sage Handbook on Environment and Society*, 2007, 239.

21. Pamela Wofford, Randy Segawa, and Jay Schreider, "Pesticide Air Monitoring in Parlier, CA," California Department of Pesticide Regulation, December 2009, i.

22. Raoul S. Liévanos, Jonathan K. London, and Julie Sze, "Uneven Transformations and Environmental Justice: Regulatory Science, Street Science, and Pesticide Regulation in California," in *Technoscience and Environmental Justice: Expert Cultures in a Grassroots Movement*, ed. Gwen Ottinger and Benjamin Cohen (Cambridge, Mass.: MIT Press, 2011), 202.

23. Liévanos, London, and Sze, 224.

24. Marc A. Edwards and Siddhartha Roy, "Academic Research in the 21st Century: Maintaining Scientific Integrity in a Climate of Perverse Incentives and Hypercompetition," *Environmental Engineering Science* 34, no. 1 (2017): 51–61, https://doi.org/10.1089/ees.2016.0223.

25. Marcia McNutt, "The New Patrons of Research," *Science* 344, no. 6179 (2014): 9.

26. Rafael Gomez and Morley Gunderson, "Volunteer Activity and the Demands of Work and Family," *Relations Industrielles / Industrial Relations* 58, no. 4 (2003): 573–589; Marc A. Musick and John Wilson, *Volunteers: A Social Profile* (Bloomington: Indiana University Press, 2007).

27. Musick and Wilson, *Volunteers*.

28. Wendy Kaminer, *Women Volunteering: The Pleasure, Pain, and Politics of Unpaid Work from 1830 to the Present* (New York: Anchor, 1984).

29. Andrea Muehlebach, *The Moral Neoliberal: Welfare and Citizenship in Italy* (Chicago: University of Chicago Press, 2012).

30. Robert N. Proctor, *Value Free Science? Purity and Power in Modern Knowledge* (Boston: Harvard University Press, 1991).

31. George H. Daniels, "The Pure-Science Ideal and Democratic Culture," *Science, New Series* 156, no. 3783 (June 30, 1967): 1699–1705.

32. Roger A. Pielke, *The Honest Broker: Making Sense of Science in Policy and Politics* (Cambridge: Cambridge University Press, 2007); Chris Mooney, *The Republican War on Science* (New York: Basic Books, 2007).

33. Nicholas Steneck, "Responsible Advocacy in Science: Standards, Benefits, and Risks," American Association for the Advancement of Science, 2011, https://www.aaas.org/resources/report-responsible-advocacy -science-standards-benefits-and-risks.

34. Gwen Ottinger, "Social Movement-Based Citizen Science," in *The Rightful Place of Science: Citizen Science*, ed. Darlene Cavalier and Eric B. Kennedy (Tempe, Ariz.: Consortium for Science, Policy, and Outcomes, 2016), 91.

35. Gili S. Drori et al., *Science in the Modern World Polity: Institutionalization and Globalization* (Stanford: Stanford University Press, 2003).

36. Daniel Lee Kleinman and Abby Kinchy, "Why Ban Bovine Growth Hormone? Science, Social Welfare, and the Divergent Biotech Policy Landscapes in Europe and the United States," *Science as Culture* 12, no. 3 (2003): 375–414.

37. Serina Rahman, *Johor's Forest City Faces Critical Challenges* (Singapore: ISEAS-Yusof Ishak Institute, 2017).

38. Linda Nash, *Inescapable Ecologies: A History of Environment, Disease, and Knowledge* (Berkeley: University of California Press, 2006).

39. Nicholas Shapiro, Nasser Zakaria, and Jody A. Roberts, "A Wary Alliance: From Enumerating the Environment to Inviting Apprehension," *Engaging Science, Technology, and Society* 3 (2017): 575–602.

40. David N. Pellow, "Popular Epidemiology and Environmental Movements: Mapping Active Narratives for Empowerment," *Humanity & Society* 21, no. 3 (1997): 319, http://has.sagepub.com/content/21/3/307 .short; Ottinger, "Social Movement-Based Citizen Science," 91.

41. Shapiro, Zakaria, and Roberts, "Wary Alliance," 580.

42. Jamie Peck and Adam Tickell, "Neoliberalizing Space," *Antipode* 34, no. 3 (June 2002): 380–404, http://www.blackwell-synergy.com/links/doi/10 .1111/1467-8330.00247; James McCarthy, "Scale, Sovereignty, and Strategy

in Environmental Governance," *Antipode* 37, no. 4 (September 2005): 731–753, https://doi.org/10.1111/j.0066-4812.2005.00523.x.

3. Investigating the Impacts of Fracking

1. Tom DiChristopher, "US Shale Oil Will Surge to Nearly 7 Million Barrels a Day in April: Dept. of Energy Forecast," *CNBC*, March 13, 2018, https://www.cnbc.com/2018/03/13/us-shale-oil-will-surge-to-nearly-7-million-barrels-a-day-in-april.html.

2. "How Much Shale Gas Is Produced in the United States?," U.S. Energy Information Administration (EIA), March 8, 2018, https://www.eia.gov/tools/faqs/faq.php?id=907&t=8.

3. Daniel J. Soeder and William M. Kappel, *Water Resources and Natural Gas Production from the Marcellus Shale*, Fact Sheet 2009-3032 (Washington, D.C.: U.S. Department of Interior, 2009).

4. "Fracking Water: It's Just So Hard to Clean," *National Geographic*, October 4, 2013, https://www.nationalgeographic.com/environment/great-energy-challenge/2013/fracking-water-its-just-so-hard-to-clean/.

5. Sally Entrekin et al., "Rapid Expansion of Natural Gas Development Poses a Threat to Surface Waters," *Frontiers in Ecology and the Environment* 9, no. 9 (November 2011): 503–511, https://doi.org/10.1890/110053; Theo Colborn et al., "Natural Gas Operations from a Public Health Perspective," *Human and Ecological Risk Assessment* 17, no. 5 (2011): 1039–1056.

6. Kelly O. Maloney et al., "Unconventional Oil and Gas Spills: Materials, Volumes, and Risks to Surface Waters in Four States of the U.S.," *Science of the Total Environment* 581/582 (2017): 369–377, https://doi.org/10.1016/j.scitotenv.2016.12.142.

7. Robert B. Jackson et al., "Increased Stray Gas Abundance in a Subset of Drinking Water Wells near Marcellus Shale Gas Extraction," *Proceedings of the National Academy of Sciences of the United States of America* 110, no. 28 (July 9, 2013): 11250–11255, https://doi.org/10.1073/pnas.1221635110; Stephen G. Osborn et al., "Methane Contamination of Drinking Water Accompanying Gas-Well Drilling and Hydraulic Fracturing," *Proceedings of the National Academy of Sciences of the United States of America* 108, no. 20 (May 17, 2011): 8172–8176, https://doi.org/10.1073/pnas.1100682108.

8. Craig Michaels, James L. Simpson, and William Wegner, *Fractured Communities: Case Studies of the Environmental Impacts of Industrial Gas Drilling* (New York: Riverkeeper, 2010), https://www.riverkeeper.org/wp -content/uploads/2010/09/Fractured-Communities-FINAL-September -2010.pdf.

9. Thomas W. Pearson, *When the Hills Are Gone: Frac Sand Mining and the Struggle for Community* (Minneapolis: University of Minnesota Press, 2017).

10. In this chapter, accounts of Pennsylvania and New York residents' per- spectives are based on research by Abby Kinchy from 2009 to 2014. This research included sixteen interviews and four focus groups with residents of four counties on either side of the New York–Pennsylvania border during the early part of the gas boom (2009–2010); a survey of watershed monitoring organizations throughout the two states in 2012; interviews with sixty volunteer researchers, professional researchers, and government scientists involved in watershed protection; and observation of multiple trainings and other meetings for volunteer watershed monitors.

11. "Pennsylvania State Energy Profile," EIA, last updated July 21, 2016, https://www.eia.gov/state/analysis.php?sid=PA#34.

12. John Hurdle and Susan Phillips, "Data Trove Offers New Details on Complaints to DEP during Shale Boom," StateImpact Pennsylvania, January 31, 2017, https://stateimpact.npr.org/pennsylvania/2017/01/31/data -trove-offers-new-details-on-complaints-to-dep-during-shale-boom/.

13. Letter from Pennsylvania DEP Northcentral Regional Office Direc- tor Nels J. Taber to private resident of Warren Center, Pennsylvania, October 13, 2010. Quoted in Abby Kinchy and Simona L. Perry, "Can Volunteers Pick Up the Slack? Efforts to Remedy Knowledge Gaps about the Watershed Impacts of Marcellus Shale Gas Development," *Duke Environmental Law and Policy Journal* 22, no. 2 (2012): 310.

14. Kirk Jalbert, Abby Kinchy, and Simona L. Perry, "Civil Society Research and Marcellus Shale Natural Gas Development: Results of a Survey of Volunteer Water Monitoring Organizations," *Journal of Environmental Studies and Sciences* 4, no. 1 (2014): 78–86, http://dx.doi.org/10.1007/s13412 -013-0155-7.

15. Abby Kinchy, Sarah Parks, and Kirk Jalbert, "Fractured Knowledge: Mapping the Gaps in Public and Private Water Monitoring Efforts in

Areas Affected by Shale Gas Development," *Environment and Planning C: Government and Policy* 34, no. 5 (2016): 879–899, https://doi.org/10 .1177/0263774X15614684.

16. A short video about the Pine Creek Water Dogs is available at https:// skytruth.org/2012/12/waterdogs/.

17. "Shale Gas Monitoring," ALLARM, n.d., http://www.dickinson.edu/ info/20173/alliance_for_aquatic_resource_monitoring_allarm/2911/ volunteer_monitoring/4.

18. Stephen M. Penningroth et al., "Community-Based Risk Assessment of Water Contamination from High-Volume Horizontal Hydraulic Fracturing," *New Solutions: A Journal of Environmental and Occupational Health Policy: NS* 23, no. 1 (January 1, 2013): 137–166, https://doi.org/10 .2190/NS.23.1.i.

19. Results of this survey are discussed in more detail in Abby Kinchy, Kirk Jalbert, and Jessica Lyons, "What Is Volunteer Water Monitoring Good For? Fracking and the Plural Logics of Participatory Science," *Political Power and Social Theory* 27 (2014): 259–289, https://doi.org/10.1108/S0198 -871920140000027017; and Jalbert, Kinchy, and Perry, "Civil Society Research."

20. Personal interview with Kinchy, August 2, 2012.

21. Candie Wilderman and Jinnieth Monismith, "Monitoring Marcellus: A Case Study of a Collaborative Volunteer Monitoring Project to Document the Impact of Unconventional Shale Gas Extraction on Small Streams," *Citizen Science: Theory and Practice* 1, no. 1 (May 20, 2016), https://doi.org/10.5334/cstp.20.

22. Gregg P. Macey et al., "Air Concentrations of Volatile Compounds near Oil and Gas Production: A Community-Based Exploratory Study," *Environmental Health* 13, no. 1 (2014): 82, https://doi.org/10.1186/1476-069X-13-82.

23. Ruth Breech et al., *Warning Signs: Toxic Air Pollution Identified at Oil and Gas Development Sites* (Battleboro, Vt.: Coming Clean, 2014), 15.

24. Macey et al., "Air Concentrations."

25. Macey et al., 15.

26. For interviews with the scientists and community organizers behind this air sampling project, see http://crowdandcloud.org/physician-diagnoses -fracking; and https://medium.com/@crowdandcloud/free-speech-in -citizen-science-q-a-with-denny-larson-f9142a6957cb.

27. Jacob Robert Matz, Sara Wylie, and Jill Kriesky, "Participatory Air Monitoring in the Midst of Uncertainty: Residents' Experiences with the Speck Sensor," *Engaging Science, Technology, and Society* 3 (2017): 464, https://doi.org/10.17351/ests2017.127. See also http://www.environmentalhealthproject.org/citizen-science-toolkit.

28. Jennifer Gabrys and Helen Pritchard, "Just Good Enough Data and Environmental Sensing: Moving beyond Regulatory Benchmarks toward Citizen Action," *International Journal of Spatial Data Infrastructures Research* 13 (2016): 4–14, https://doi.org/10.2902/1725-0463.2018.13.art2. See the Data Stories on the project website, https://citizensense.net/data-stories-pa/#intro.

29. Natasha Vicens, "How One Resident near Fracking Got the EPA to Pay Attention to Her Air Quality," *PublicSource*, December 15, 2016, https://www.publicsource.org/how-one-resident-near-fracking-got-the-epa-to-pay-attention-to-her-air-quality/.

30. Sara Wylie et al., "Materializing Exposure: Developing an Indexical Method to Visualize Health Hazards Related to Fossil Fuel Extraction," *Engaging Science, Technology, and Society* 3 (2017): 426, https://doi.org/10.17351/ests2017.123. For images of the testing tools and results, see https://sarawylie.com/2014/08/01/hydrogen-sulfide-sensing/.

31. Sara Ann Wylie, *Fractivism: Corporate Bodies and Chemical Bonds* (Durham, N.C.: Duke University Press, 2018), 165–190.

32. Though WellWatch no longer exists, documentation can be found on Wylie's website, https://sarawylie.com/publications/fractivism-corporate-bodies-and-chemical-bonds/wellwatch-appendix-for-fractivism/.

33. Wylie et al., "Materializing Exposure."

34. Sam Rubight, "Help Us to Track Oil Trains in Pittsburgh and Beyond," FracTracker Alliance, October 15, 2014, https://www.fractracker.org/2014/10/track-oil-trains/.

35. Ashley Ahearn, "Washington State's Oil Train Traffic Is Shrouded in Secrecy," KUOW, June 16, 2015, http://kuow.org/post/washington-states-oil-train-traffic-shrouded-secrecy.

36. "25 Million Live in Oil Train Blast Zone: New Online Mapping Tool Shows Threat to Homes, Schools, and Cities," Forest Ethics, n.d., http://drcinfo.org/2014/09/22/25-million-live-in-oil-train-blast-zone-new-online-mapping-tool-shows-threat-to-homes-schools-and-cities-forest-ethics/.

37. Deborah Thomas, "Living with Oil and Gas and Practicing Community Conducted Science," *Engaging Science, Technology, and Society* 3 (2017): 615–616, https://doi.org/10.17351/ests2017.131.

38. U.S. Environmental Protection Agency, *Surface Water Monitoring: A Framework for Change* (Washington, D.C.: USEPA, 1987).

39. USEPA, *Volunteer Water Monitoring: A Guide for State Managers* (Washington, D.C.: USEPA, 1990).

40. Max Pfeffer and Linda Plummer Wagenet, "Volunteer Environmental Monitoring, Knowledge Creation and Citizen–Scientist Interaction," in *Sage Handbook on Environment and Society*, 2007, 235–249.

41. Diane Wilson, "Community Based Water Monitoring and beyond, a Case Study: Pennsylvania," *Proceedings of the Water Environment Federation* (2002): 1025–1036, http://www.ingentaconnect.com/content/wef/wefproc/2002/00002002/00000002/art00063.

42. Personal interview with Kinchy, April 11, 2012.

43. William Deutsch, Laura Lhotka, and Sergio Ruiz-Cordova, "Group Dynamics and Resource Availability of a Long-Term Volunteer Water-Monitoring Program," *Society & Natural Resources* 22, no. 7 (2009): 647, http://www.tandfonline.com/doi/abs/10.1080/08941920802078216.

44. Wilderman and Monismith, "Monitoring Marcellus."

45. "Fake Science: SWPA Enviro Health Registry for Those near Fracking," *Marcellus Drilling News*, May 1, 2017, http://marcellusdrilling.com/2017/05/fake-science-swpa-enviro-health-registry-for-those-near-fracking/.

46. In contrast, it should be noted that scientists who work with the industry tend to avoid such accusations and retain an untarnished image. See chapter 10 in Wylie, *Fractivism*.

47. Laura Legere, "Citizen Training for Spotting Drilling Problems Criticized by Natural Gas Industry," *Scranton Times-Tribune*, December 1, 2009.

48. Gwen Ottinger, "Social Movement-Based Citizen Science," in *The Rightful Place of Science: Citizen Science*, ed. Darlene Cavalier and Eric B. Kennedy (Tempe, Ariz.: Consortium for Science, Policy, and Outcomes, 2016), 98.

49. Katie Brown, "New Air Quality Report Uses Scientifically Dubious Methods," Energy in Depth, October 30, 2014, https://energyindepth.org/national/new-air-quality-report-uses-scientifically-dubious-methods/.

50. Thanks to Sara Wylie for pointing this out to us.

51. Simona L. Perry, "Development, Land Use, and Collective Trauma: The Marcellus Shale Gas Boom in Rural Pennsylvania," *Culture, Agriculture, Food and Environment* 34, no. 1 (June 26, 2012): 81–92, https://doi.org/10.1111/j.2153-9561.2012.01066.x; Kathryn J. Brasier et al., "Risk Perceptions of Natural Gas Development in the Marcellus Shale," *Environmental Practice* 15, no. 2 (2013): 108–122.

52. Emily Eaton and Abby Kinchy, "Quiet Voices in the Fracking Debate: Ambivalence, Nonmobilization, and Individual Action in Two Extractive Communities (Saskatchewan and Pennsylvania)," *Energy Research and Social Science* 20 (2016): 22–30, https://doi.org/10.1016/j.erss.2016.05.005.

53. Statement by Bradford County, Pennsylvania, resident during focus group discussion moderated by Kinchy, October 20, 2010.

54. Ann M. Eisenberg, "Beyond Science and Hysteria: Reality and Perceptions of Environmental Justice Concerns Surrounding Marcellus and Utica Shale Gas Development," *University of Pittsburgh Law Review* 77 (2015), https://doi.org/10.5195/lawreview.2015.396.

55. Stephanie A. Malin and Kathryn Teigen DeMaster, "A Devil's Bargain: Rural Environmental Injustices and Hydraulic Fracturing on Pennsylvania's Farms," *Journal of Rural Studies* 47, no. 2015 (2016): 278–290, https://doi.org/10.1016/j.jrurstud.2015.12.015.

56. Ken Dufalla, "Bromides: Where Forth They Come," *Greene County Messenger*, March 2, 2012, http://www.heraldstandard.com/gcm/columns/natures_corner/bromides-where-forth-they-come/article_29d6632a-60e0-5c47-919c-3a0a05e2d105.

57. Andrew Stacy, "Additional WVU Testing Confirms Acceptable Levels of Radioactivity in Drinking Water at Clyde Mine," *Three Rivers Quest News*, August 27, 2015, http://3riversquest.org/additional-wvu-testing-confirms-radioactivity-below-drinking-water-standards-at-clyde-mine/; Bob Niedbala, "DEP Final Report on Ten Mile Creek Indicates No Dangers of Radioactivity," *Observer-Reporter* (Washington, Pa.), December 2, 2016, http://www.observer-reporter.com/20161202/dep_final_report_on_ten_mile_creek_indicates_no_dangers_of_radioactivity.

58. Natasha Vicens, "DEP's Testing Methods for Radiation in a PA Creek Questioned," *PublicSource*, July 30, 2015, http://publicsource.org/deps-testing-methods-for-radiation-in-a-pa-creek-questioned/.

59. Natasha Vicens, "PA DEP Finds Safe Radioactivity Levels in Greene County Creek," *PublicSource*, December 17, 2015, http://publicsource.org/pa-dep-finds-safe-radioactivity-levels-in-greene-county-creek/.

60. "Clyde Mine Discharge / Ten Mile Creek Water Quality Final Report," Pennsylvania Department of Environmental Protection, November 1, 2016, https://www.slideshare.net/MarcellusDN/clyde-mine-dischargetenmile-creek-water-quality-final-report.

61. Skylar Zilliox and Jessica M. Smith, "Colorado's Fracking Debates: Citizen Science, Conflict and Collaboration," *Science as Culture* 5431 (2018): 1–21, https://doi.org/10.1080/09505431.2018.1425384.

62. Zilliox and Smith, 224.

63. S. L. Brantley et al., "Engaging over Data on Fracking and Water Quality: Data Alone Aren't the Solution, but They Bring People Together," *Science* 359, no. 6374 (2018): 397, https://doi.org/10.1126/science.aan6520; Maloney et al., "Unconventional Oil and Gas Spills."

4. Detecting Radiation

1. In this chapter, accounts of CRMOs are based on research by Kimura from 2011 to 2017. A national list of seventy-four active CRMOs was created based on media accounts and websites of CRMOs. Of these, sixty-five were interviewed either in person or on the phone, mostly in 2013–2014, with follow-up interviews with a number of them until 2017. In addition, interviews with researchers and government officials, media reports, government documents were used in this analysis.

2. "More and More Food Found above Standards: Producers Say Standards Too Strict," *Yomiuri Shinbun Newspaper*, March 2011.

3. Agency for Natural Resources and Energy, "Wagakuni niokeru genshiryokuseisakuno genjo" [Current status of nuclear reactors in Japan], Enecho.meti, 2018, http://www.enecho.meti.go.jp/category/electricity_and_gas/nuclear/001/pdf/001_02_001.pdf.

4. Gabrielle Hecht, *Being Nuclear: Africans and the Global Uranium Trade* (Cambridge, Mass.: MIT Press, 2012).

5. Kristin Shrader-Frechette, *What Will Work: Fighting Climate Change with Renewable Energy, Not Nuclear Power* (Oxford: Oxford University

Press, 2011); Tadashi Yagi, *Genpatsuwa Sabetsude Ugoku* [Nuclear energy is rooted in discrimination] (Tokyo: Akashi shoten, 2011).

6. For instance, TEPCO admitted in 2002 that it did not report unfavorable accident data to the government as mandated by law. TEPCO also admitted that it had falsified data in two hundred cases, but the government's oversight did not lead to overhauling the safety systems. In 2007, the company revealed yet more unreported accidents. Hiroaki Koide, *Genpatsu no uso* [Lies of the nuclear power plants] (Tokyo: Fusosha, 2011).

7. Minato Kawamura, *Fukushimagenpatsu jinsaiki* [Fukushima nuclear reactor human disaster] (Tokyo: Gendaishokan, 2011), 90.

8. Aya H. Kimura, *Radiation Brain Moms and Citizen Scientists: The Gender Politics of Food Contamination after Fukushima* (Durham, N.C.: Duke University Press, 2016).

9. David H. Slater, Rika Morioka, and Haruka Danzuka, "Micro-politics of Radiation," *Critical Asian Studies* 46, no. 3 (July 2014): 485–508.

10. Kimura, *Radiation Brain Moms.*

11. See the Ministry of Agriculture, Forestry, and Fisheries website at http://www.maff.go.jp/j/kanbo/joho/saigai/s_chosa/H25gaiyo.html. The seafood data are calculated based on a report by the Fisheries Agency. Their reports are available at http://www.jfa.maff.go.jp/j/housyanou/kekka.html.

12. Kimura, *Radiation Brain Moms.*

13. Atsuro Morita, Anders Blok, and Shuhei Kimura, "Environmental Infrastructures of Emergency: The Formation of a Civic Radiation Monitoring Map during the Fukushima Disaster," in *Nuclear Disaster at Fukushima Daiichi: Social, Political and Environmental Issues*, ed. Richard Hindmarsh (New York: Routledge, 2013), 78–96; Jean-Christophe Plantin, "The Politics of Mapping Platforms: Participatory Radiation Mapping after the Fukushima Daiichi Disaster," *Media, Culture & Society* 37, no. 6 (2015): 904–921.

14. The website can be seen at http://safecast.org/tilemap/.

15. Keisuke Katsuki, "Hosyanosokutei purojekuto Safecast" [Radiation measurement project, Safecast], *Politas*, March 2015, http://politas.jp/features/4/article/354; https://blog.safecast.org/ja/.

16. Azby Brown et al., "Safecast: Successful Citizen-Science for Radiation Measurement and Communication after Fukushima," *Journal of Radiological Protection* 36, no. 2 (2016): S17, https://doi.org/10.1088/0952-4746/36/2/S82.

17. The project website can be seen at https://minnanods.net/soil/.

18. The data can be viewed at project websites https://fukushimainform.ca and http://ourradioactiveocean.org.

19. John N. Smith et al., "Recent Transport History of Fukushima Radioactivity in the Northeast Pacific Ocean," *Environmental Science and Technology* 51, no. 18 (2017): 10494–10502, https://doi.org/10.1021/acs.est.7b02712.

20. We thank Gabrielle Hecht for this insight.

21. Akiko Hemmi and Ian Graham, "Hacker Science versus Closed Science: Building Environmental Monitoring Infrastructure," *Information, Communication & Society* 17, no. 7 (2014): 830, https://doi.org/10.1080/1369118X.2013.848918.

22. Yasuhito Abe, "Safecast, or The Production of Collective Intelligence on Radiation Risks after 3.11," *Asia-Pacific Journal* 12, no. 7 (2014): 1–11.

23. Personal interview with Kimura, December 24, 2013.

24. Personal interview with Kimura, February 24, 2014.

25. Patricia Steinhoff, "Memories of New Left Protest," *Journal of the German Institute for Japanese* 25, no. 2 (2013): 127–165.

26. Harutoshi Funabashi, Nobuko Iijima, and Koichi Hasegawa, *Kakunenryōsaikurushisetsuno shakaigaku* [Sociology of nuclear fuel recycle facilities] (Tokyo: Yuhikaku, 2012); Yoshinari Ichinomiya et al., *Genpatsusaikado no fukaiyami* [Problems of restarting the nuclear reactors] (Tokyo: Takarajimasha, 2012); Kenji Higuchi, *Yaminikesareru genpatuhibakusya* [Invisible exposed people] (Tokyo: Hachigatsu Shobo, 2011); Yagi, *Genpatsuwa sabetsude ugoku.*

27. "Hōshasen yori hinanrisuku ga shinpai" [More concerns about risk of evacuation than radiation], *Sankei Shinbun Newspaper*, May 26, 2015, http://www.sankeibiz.jp/compliance/news/160526/cpd1605260500002-n5.htm.

28. Hikaru Amano et al., "Method for Rapid Screening Analysis of Sr-90 in Edible Plant Samples Collected near Fukushima, Japan," *Applied Radiation and Isotopes* 112 (2016): 131–135, https://doi.org/10.1016/j.apradiso.2016.03.026.

29. Tetsunari Iida, Eisaku Sato, and Taro Kono, *Genshiryokumura wo koete* [Beyond the nuclear village] (Tokyo: NHK Books, 2011).

30. Tadashi Kobayashi, *Toransu saiensu no jidai* [The age of trans-science] (Tokyo: NTT Shuppan, 2007), chap. 4.

31. Jinzaburo Takagi, *Shimin kagakushatoshite ikiru* [Life as a citizen scientist], vol. 1 (Tokyo: Nanatsumorishokan, 2002), 68.

32. Takagi, 72.

33. Hiroaki Koide, *Hōshanoosen no genjitsu wo koete* [Beyond the reality of radiation contamination] (Tokyo: Hokuto Shuppan, 1992), 14.

34. Aya H. Kimura, "Risk Communication under Post-feminism: Analysis of Risk Communication Programmes after the Fukushima Nuclear Accident," *Science Technology & Society* 21, no. 1 (2016): 24–41, https://doi .org/10.1177/0971721815622738; Aya H. Kimura, "Fukushima ETHOS: Post-disaster Risk Communication, Affect, and Shifting Risks," *Science as Culture* 27, no. 1 (2018): 98–117, https://doi.org/10.1080/09505431.2017 .1325458.

35. "Josenkijyun kanwa" [Relaxing decontamination standards], Ourplanet, August 2014, http://www.ourplanet-tv.org/?q=node/1814.

36. Masakazu Sugiura, "Heisei 26 nendo Kankyōshō oyobi genshiryokuki-seikankeiyosan no gaiyo" [Overview of budget for the Ministry of Environment and Nuclear Power Regulations for FY 2014], *Rippo to chosa* 349 (2014): 134–136.

37. *Kojinhibakusenryōsokutei no jireisyu* [Examples of personal exposure monitoring], report, team in charge of assisting the lives of disaster victims, 2015, http://www.meti.go.jp/earthquake/nuclear/kinkyu/committee/ advisor/2015/pdf/0826_01e.pdf.

38. *Kikannimuketa anzen anshin taisakuni kansurukihontekikangaekata* [Basic concept of safety and security measures for evacuees to return home], Nuclear Regulation Authority, November 2014.

39. Kimura, "Fukushima ETHOS."

40. Hideyuki Hirakawa and Masashi Shirabe, "Rhetorical Marginalization of Science and Democracy: Politics in Risk Discourse on Radioactive Risks in Japan," in *Lessons from Fukushima: Japanese Case Studies on Science, Technology and Society*, ed. Yuko Fujigaki (London: Springer International, 2015), 57–86, https://doi.org/10.1007/978-3-319-15353-7_4. Also see Olga Kuchinskaya, *The Politics of Invisibility: Public Knowledge*

about Radiation Health Effects after Chernobyl (Cambridge, Mass.: MIT Press, 2014).

41. Specifically, critics have argued that the glass badge is designed to be used in situations of one-directional exposure to radiation. When used daily and with radiation coming from all directions, the glass badge will underestimate dosage. Furthermore, although the user is supposed to carry it around all the time, some people might not wear the dosimeter but leave it in a bag or somewhere inside, which would lower the reading. The D-Shuttle entails an additional concern of privacy violation, as it shows hour-by-hour readings. Most profoundly, critics have pointed out that regulations on radiological exposure for workers (such as in the Ordinance on Prevention of Ionizing Radiation Hazards) have a two-layered mechanism of radiological protection that combines control by space and control by person, but the current policy would mean providing general citizens less protection because it involves control only by person and not by space. See Kazumasa Aoki, *Kōjōsenkensawo meguru senmonkano giron* [Expert discussions on thyroid screening], Foejapan.org, January 2015, http://www.foejapan.org/energy/evt/150118_aoki.pdf; "Garasubajji" [Glass badge], Fukushimarokyugenpatsuwo kangaeru kai, January 2015, http://fukurou.txt-nifty.com/fukurou/2015/01/post-156b.html.

42. We thank Gabrielle Hecht for this insight.

43. Allen H. Barton, *Communities in Disaster: A Sociological Analysis of Collective Stress Situations* (Garden City, N.Y.: Doubleday, 1969).

44. Ben Wisner et al., *At Risk: Natural Hazards, People's Vulnerability and Disasters* (London: Routledge, 2004); Charles Perrow, *Normal Accidents: Living with High-Risk Technologies* (New York: Basic Books, 1984); E. L. Quarantelli, ed., *What Is a Disaster? Perspectives on the Question* (New York: Taylor & Francis, 1998).

5. Tracking Genetically Engineered Crops

1. David J. Hess, "The Potentials and Limitations of Civil Society Research: Getting Undone Science Done," *Sociological Inquiry* 79, no. 3 (August 2009): 306–327, doi:10.1111/j.1475-682X.2009.00292.x; Lara Khoury and Stuart Smyth, "Reasonable Foreseeability and Liability in Relation to Genetically Modified Organisms," *Bulletin of Science,*

Technology & Society 27, no. 3 (2007): 215–232, https://doi.org/10.1177/0270467607300635; Michael R. Taylor and Jody S. Tick, *The StarLink Case: Issues for the Future* (Washington, D.C.: Resources for the Future / Pew Initiative on Food and Biotechnology, 2001); ETC Group, "Nine Mexican States Found to Be GM Contaminated" (news release, 2003), http://www.etcgroup.org/en/materials/publications.html?id=145.

2. Aventis was later acquired by Bayer, one of the five biggest agrochemical and seed companies in the world.

3. Robert Gottlieb and Anupama Joshi, *Food Justice* (Cambridge, Mass.: MIT Press, 2010), 6.

4. There are many excellent studies on the impacts of multinational agrochemical corporations' control over agrifood systems, such as Jack Kloppenburg, *First the Seed: The Political Economy of Plant Biotechnology* (Madison: University of Wisconsin Press, 2005); Vandana Shiva, *Monocultures of the Mind: Perspectives on Biodiversity and Biotechnology* (London: Zed Books, 1993); and Rachel A. Schurman and Dennis Doyle Takahashi Kelso, *Engineering Trouble: Biotechnology and Its Discontents*, 1st ed. (Berkeley: University of California Press, 2003).

5. This section is based on Kinchy's previous research. Abby Kinchy, *Seeds, Science, and Struggle: The Global Politics of Transgenic Crops* (Cambridge, Mass.: MIT Press, 2012). Data was collected through interviews, participant observations, archival research, and analyses of published documents such as legal and scientific papers, newspaper articles, and activist websites. Field research in Mexico took place from September 2005 to April 2006. Interviews were conducted in Mexico City and the state of Oaxaca with more than fifty activists, farmers, intellectuals, researchers, plant breeders, and government officials involved in the debate over GE corn. In addition, observations were made in a wide array of settings, including meetings organized by federal environmental agencies, activist meetings, and public demonstrations.

6. Gustavo Esteva and Catherine Marielle, *Sin maíz no hay país* (Mexico City: Museo Nacional de Culturas Populares, 2003); Roberto J. González, *Zapotec Science: Farming and Food in the Northern Sierra of Oaxaca* (Austin: University of Texas Press, 2001).

7. Secretariat of the Commission for Environmental Cooperation, *Maize & Biodiversity: The Effects of Transgenic Maize in Mexico*

(Montreal: Communications Department of the Commission for Environmental Cooperation, 2004).

8. Gisele Henriques and Raj Patel, *Policy Brief No. 7: Agricultural Trade Liberalization and Mexico* (Oakland, Calif.: Food First / Institute for Food and Development Policy, 2003), 24.

9. Mauricio R. Bellon and Julien Berthaud, "Traditional Mexican Agricultural Systems and the Potential Impacts of Transgenic Varieties on Maize Diversity," *Agriculture and Human Values* 23 (2006): 3–14.

10. Tom Barry, *Zapata's Revenge: Free Trade and the Farm Crisis in Mexico* (Boston: South End Press, 1995).

11. Elizabeth Fitting, "Importing Corn, Exporting Labor: The Neoliberal Corn Regime, GMOs, and the Erosion of Mexican Biodiversity," *Agriculture and Human Values* 23, no. 1 (2006): 15–26.

12. Annette-Aurélie Desmarais, "The Vía Campesina: Consolidating an International Peasant and Farm Movement," *Journal of Peasant Studies* 29, no. 2 (2002): 91–124.

13. "GREENPEACE Confirms USA Is Introducing Transgenic Maize into Mexico" (press release, May 25, 1999), archived by Genet News listserv, http://www.gene.ch/genet/1999/May/msg00104.html.

14. David Quist and Ignacio H. Chapela, "Transgenic DNA Introgressed into Traditional Maize Landraces in Oaxaca, Mexico," *Nature* 414, no. 6863 (2001): 541–543.

15. Jason A. Delborne, "Transgenes and Transgressions: Scientific Dissent as Heterogeneous Practice," *Social Studies of Science* 38, no. 4 (2008): 509–541, https://doi.org/Doi10.1177/0306312708089716.

16. Exequiel Ezcurra and Jorge Soberón Mainero, "Evidence of Gene Flow from Transgenic Maize to Local Varieties in Mexico," in *LMOs and the Environment: Proceedings of an International Conference*, ed. OECD (Paris: OECD, 2002), 289–295.

17. Kinchy, *Seeds, Science, and Struggle*, 82.

18. Personal interview with Kinchy, December 9, 2005.

19. Kinchy, 83–84.

20. Kinchy, 84.

21. David Alire Garcia, "Monsanto Sees Prolonged Delay on GMO Corn Permits in Mexico," *Reuters*, January 20, 2017, http://www.reuters.com/article/us-mexico-monsanto-idUSKBN15E1DJ.

22. The information on this case is drawn from multiple sources. Kimura interviewed a key organizer of the national network of anti-GMO groups as well as an organizer of a citizen science group in the western part of Japan in July 2017 and another in May 2018. She also attended a national meeting where all groups reported on their findings from their citizen science projects in regards to GE rapeseed in 2017. In addition, she obtained reports from past studies that summarized findings of several citizens' groups.

23. John M. Pleasants and Karen S. Oberhauser, "Milkweed Loss in Agricultural Fields Because of Herbicide Use: Effect on the Monarch Butterfly Population," *Insect Conservation and Diversity* 6, no. 2 (March 2013): 135–144, https://doi.org/10.1111/j.1752-4598.2012.00196.x.

24. Pleasants and Oberhauser, 135–144.

25. John M. Pleasants et al., "Interpreting Surveys to Estimate the Size of the Monarch Butterfly Population: Pitfalls and Prospects," ed. Travis Longcore, *PLOS ONE* 12, no. 7 (July 14, 2017): 1, https://doi.org/10.1371/journal.pone.0181245.

26. Pleasants et al., 14.

27. "Biofortified Consulting," Biology Fortified, n.d., https://www.biofortified.org/projects/consulting/.

28. Karl Haro von Mogel, "Final Steps on the GMO Corn Experiment," Biology Fortified, data from August 20, 2017, https://www.biofortified.org/2017/08/final-steps-gmo-experiment/.

29. Karl Haro von Mogel and Anastasia Bodnar, "The GMO Corn Experiment," Experiment.com, data from March 30, 2018, https://experiment.com/projects/the-gmo-corn-experiment.

30. "Scientists Launch a Citizen Science Experiment on GMOs," Biology Fortified, October 28, 2015, https://www.biofortified.org/2015/10/scientists-launch-a-citizen-science-experiment-on-gmos/.

31. Biofortified Board, "Statement on Kevin Folta and Conflicts of Interest," Biology Fortified, August 29, 2018, https://biofortified.org/2018/08/kevin-folta-coi/.

32. "Our Mission," Cornell Alliance for Science, n.d., https://allianceforscience.cornell.edu/about/mission/.

33. "Calling All Citizen Scientists: Help Evaluate GMO Peer-Reviewed Literature!," Cornell Alliance for Science, July 12, 2016, http://

allianceforscience.cornell.edu/blog/calling-all-citizen-scientists-help
-evaluate-gmo-peer-reviewed-literature.

34. Michael N. Antoniou and Claire J. Robinson, "Cornell Alliance for
Science Evaluation of Consensus on Genetically Modified Food Safety:
Weaknesses in Study Design," *Frontiers in Public Health* 5 (2017): 79,
https://doi.org/10.3389/fpubh.2017.00079.

35. Personal interview with Kimura, July 8, 2017.

36. Eva J. Lewandowski and Karen S. Oberhauser, "Butterfly Citizen Science Projects Support Conservation Activities among Their Volunteers,"
Citizen Science: Theory and Practice 1, no. 6 (2016): 1–8, https://doi.org/10
.5334/cstp.10.

37. Lewandowski and Oberhauser.

38. Charles M. Benbrook, "Trends in Glyphosate Herbicide Use in the
United States and Globally," *Environmental Sciences Europe* 28, no. 1
(2016): 3, https://doi.org/10.1186/s12302-016-0070-0.

39. Alexis Baden-Mayer, "Time to Ban Monsanto's Roundup
Weedkiller—but Will the EPA Act?," *Organic Consumer* (blog), April 22,
2018, https://www.organicconsumers.org/blog/time-ban-monsantos
-roundup-weedkiller-will-epa-act.

40. "Laureates Letter Supporting Precision Agriculture (GMOs)," Support
Precision Agriculture, June 29, 2016, http://supportprecisionagriculture
.org/nobel-laureate-gmo-letter_rjr.html.

41. "Golden Rice," Greenpeace, 2013, http://www.greenpeace.org/
international/en/campaigns/agriculture/problem/Greenpeace-and
-Golden-Rice/.

42. Aya H. Kimura, *Hidden Hunger: Gender and the Politics of Smarter Foods*
(Ithaca, N.Y.: Cornell University Press, 2013).

43. For a detailed account of this debate within the activist network, see
Kinchy, *Seeds, Science, and Struggle*, 87–95.

44. Daniel Lee Kleinman and Abby Kinchy, "Why Ban Bovine Growth
Hormone? Science, Social Welfare, and the Divergent Biotech Policy
Landscapes in Europe and the United States," *Science as Culture* 12, no. 3
(2003): 375–414; Daniel Lee Kleinman and Abby Kinchy, "Against the
Neoliberal Steamroller? The Biosafety Protocol and the Social Regulation of Agricultural Biotechnologies," *Agriculture and Human Values* 24,
no. 2 (2007): 195–206.

45. Personal interview with Kimura, July 8, 2017.

46. Although safety approvals were granted to eight crops, only GE roses are commercially harvested because of consumer and farmer opposition.

47. Taylor and Tick, *StarLink Case*.

48. Philip H. Howard, "Intellectual Property and Consolidation in the Seed Industry," *Crop Science* 55, no. 6 (2015): 2489–2495.

Conclusion

1. Jon D. Wisman, "Inequality, Social Respectability, Political Power, and Environmental Devastation," *Journal of Economic Issues* 45, no. 4 (2011): 877–900.

2. Hans Radder, *The Commodification of Academic Research: Science and the Modern University* (Pittsburgh: Pittsburgh University Press, 2010); Daniel Lee Kleinman, *Impure Cultures: University Biology and the World of Commerce* (Madison: University of Wisconsin Press, 2003).

3. Muki Haklay, *Citizen Science and Policy: A European Perspective*, vol. 4 (Washington, D.C.: Woodrow Wilson Center, 2015).

4. For a detailed summary of issues and ideas identified by this working group, see Daniela Soleri et al., "Finding Pathways to More Equitable and Meaningful Public-Scientist Partnerships," *Citizen Science: Theory and Practice* 1, no. 1 (2016): 1–11, http://doi.org/10.5334/cstp.46. See also Norman Porticella et al., *Promising Practices for Community Partnerships: A Call to Support More Inclusive Approaches to Public Participation in Scientific Research* (Washington, D.C.: Association of Science-Technology Centers, 2013).

5. Barbara L. Allen, "Strongly Participatory Science and Knowledge Justice in an Environmentally Contested Region," *Science, Technology & Human Values* 43, no. 6 (2018): 949, https://doi.org/10.1177/0162243918758380.

6. It is noteworthy that many religious conservative groups deny mainstream scientific ideas only for some key issues (for instance, evolution, homosexuality, climate change, and stem cell research), but even then, they argue for the "correct" science that aligns with their views. Antony Alumkal, *Paranoid Science: The Christian Right's War on Reality* (New York: New York University Press, 2017).

7. "Safecast Data FAQ," Safecast blog, n.d., https://blog.safecast.org/faq/ data/.

8. Gwen Ottinger, "Is It Good Science? Activism, Values, and Communicating Politically Relevant Science," *JCOM—Journal of Science Communication* 14, no. 2 (2015): C02.

9. The March for Science also prompted the rebirth of Science for the People, a radical scientists' organization that formed in the 1960s to protest research investments in militarism, industrial interests, and environmental exploitation. Today's Science for the People is made up of "STEM workers, educators, and activists who believe that science can be a positive force for humanity and the planet." These ideas are consistent with the vision we are articulating here (https://scienceforthepeople .org/).

10. Sainath Suryanarayanan and Daniel Lee Kleinman, *Vanishing Bees: Science, Politics, and Honeybee Health* (New Brunswick, N.J.: Rutgers University Press, 2016); Sainath Suryanarayanan et al., "Collaboration Matters: Honey Bee Health as a Transdisciplinary Model for Understanding Real-World Complexity," *BioScience* 68, no. 12 (2018): 990–995.

Bibliography

Abe, Yasuhito. "Safecast, or The Production of Collective Intelligence on Radiation Risks after 3.11." *Asia-Pacific Journal* 12, no. 7 (2014): 1–11.

Agency for Natural Resources and Energy. "Wagakuni niokeru genshiryoku-seisakuno genjo" [Current status of nuclear reactors in Japan]. Enecho. meti, 2018. http://www.enecho.meti.go.jp/category/electricity_and_gas/nuclear/001/pdf/001_02_001.pdf.

Agrawal, Arun, and Clark C. Gibson. *Communities and the Environment: Ethnicity, Gender, and the State in Community-Based Conservation.* New Brunswick, N.J.: Rutgers University Press, 2001.

Ahearn, Ashley. "Washington State's Oil Train Traffic Is Shrouded in Secrecy." KUOW, June 16, 2015. http://kuow.org/post/washington-states-oil-train-traffic-shrouded-secrecy.

Alire Garcia, David. "Monsanto Sees Prolonged Delay on GMO Corn Permits in Mexico." *Reuters,* January 20, 2017. http://www.reuters.com/article/us-mexico-monsanto-idUSKBN15E1DJ.

Allen, Barbara L. "Strongly Participatory Science and Knowledge Justice in an Environmentally Contested Region." *Science, Technology & Human Values* 43, no. 6 (2018): 947–971. https://doi.org/10.1177/0162243918758380.

———. *Uneasy Alchemy: Citizens and Experts in Louisiana's Chemical Corridor Disputes.* Cambridge, Mass.: MIT Press, 2003.

Alumkal, Antony. *Paranoid Science: The Christian Right's War on Reality.* New York: New York University Press, 2017.

Amano, Hikaru, Hideaki Sakamoto, Norikatsu Shiga, and Kaori Suzuki. "Method for Rapid Screening Analysis of Sr-90 in Edible Plant Samples Collected near Fukushima, Japan." *Applied Radiation and Isotopes* 112 (2016): 131–135. https://doi.org/10.1016/j.apradiso.2016.03.026.

Antoniou, Michael N., and Claire J. Robinson. "Cornell Alliance for Science Evaluation of Consensus on Genetically Modified Food Safety: Weaknesses in Study Design." *Frontiers in Public Health* 5 (2017): 79. https://doi.org/10.3389/fpubh.2017.00079.

Aoki, Kazumasa. *Kōjōsenkensawo meguru senmonkano giron* [Expert discussions on thyroid screening]. Foejapan.org, January 2015. http://www.foejapan.org/energy/evt/150118_aoki.pdf.

Baden-Mayer, Alexis. "Time to Ban Monsanto's Roundup Weedkiller—but Will the EPA Act?" *Organic Consumer* (blog), April 22, 2018. https://www.organicconsumers.org/blog/time-ban-monsantos-roundup-weedkiller-will-epa-act.

Ballard, Heidi L., and Brinda Sarathy. "Inclusion and Exclusion: Immigrant Forest Workers and Participation in Natural Resource Management." In *Partnerships for Empowerment: Participatory Research for Community-Based Natural Resource Management*, edited by Carl Wilmsen, William Elmendorf, Larry Fisher, Jacquelyn Ross, Brinda Sarathy, and Gail Wells, 187–216. London: Earthscan, 2008.

Barry, Tom. *Zapata's Revenge: Free Trade and the Farm Crisis in Mexico*. Boston: South End Press, 1995.

Barton, Allen H. *Communities in Disaster: A Sociological Analysis of Collective Stress Situations*. Garden City, N.Y.: Doubleday, 1969.

Bellon, Mauricio R., and Julien Berthaud. "Traditional Mexican Agricultural Systems and the Potential Impacts of Transgenic Varieties on Maize Diversity." *Agriculture and Human Values* 23 (2006): 3–14.

Benbrook, Charles M. "Trends in Glyphosate Herbicide Use in the United States and Globally." *Environmental Sciences Europe* 28, no. 1 (2016): 3. https://doi.org/10.1186/s12302-016-0070-0.

Bidwell, David. "Is Community-Based Participatory Research Postnormal Science?" *Science, Technology & Human Values* 34, no. 6 (2009): 741–761. https://doi.org/10.1177/0162243909340262.

Biofortified Board. "Statement on Kevin Folta and Conflicts of Interest." Biology Fortified, August 29, 2018. https://biofortified.org/2018/08/kevin-folta-coi/.

Bonney, Rick, Jennifer Shirk, and Tina B. Philip. "Citizen Science." In *Encyclopedia of Science Education*, edited by Richard Gunstone, 152–154. Dordrecht: Springer, 2015.

Bonney, Rick, Jennifer Shirk, Tina B. Phillips, Andrea Wiggins, Heidi L. Ballard, Abraham J. Miller-Rushing, and Julia K. Parrish. "Citizen Science: Next Steps for Citizen Science." *Science* 343, no. 6178 (2014): 1436–1437. https://doi.org/10.1126/science.1251554.

Brantley, S. L., R. D. Vidic, K. Brasier, D. Yoxtheimer, J. Pollak, C. Wilderman, and T. Wen. "Engaging over Data on Fracking and Water Quality: Data Alone Aren't the Solution, but They Bring People Together." *Science* 359, no. 6374 (2018): 395–397. https://doi.org/10.1126/science.aan6520.

Brasier, Kathryn J., Diane K. McLaughlin, Danielle Rhubart, Richard C. Stedman, Matthew R. Filteau, and Jeffrey Jacquet. "Risk Perceptions of Natural Gas Development in the Marcellus Shale." *Environmental Practice* 15, no. 2 (2013): 108–122.

Breech, Ruth, Caroline Cox, Elizabeth Crowe, Jessica Hendricks, and Denny Larson. *Warning Signs: Toxic Air Pollution Identified at Oil and Gas Development Sites*. Battleboro, Vt.: Coming Clean, 2014.

Breen, Jessica, Shannon Dosemagen, Jeffrey Warren, and Mathew Lippincott. "Mapping Grassroots: Geodata and the Structure of Community-Led Open Environmental Science." *ACME: An International E-journal for Critical Geographies* 14, no. 3 (2015): 849–873.

Brown, Azby, Pieter Franken, Sean Bonner, Nick Dolezal, and Joe Moross. "Safecast: Successful Citizen-Science for Radiation Measurement and Communication after Fukushima." *Journal of Radiological Protection* 36, no. 2 (2016): S82–S101. https://doi.org/10.1088/0952-4746/36/2/S82.

Brown, Katie. "New Air Quality Report Uses Scientifically Dubious Methods." Energy in Depth, October 30, 2014. https://energyindepth.org/national/new-air-quality-report-uses-scientifically-dubious-methods/.

Brown, Phil. *Toxic Exposures: Contested Illnesses and the Environmental Health Movement*. New York: Columbia University Press, 2007.

———. "When the Public Knows Better: Popular Epidemiology." *Environment* 35, no. 8 (1993): 16–21.

Bryan, Joe. "Walking the Line: Participatory Mapping, Indigenous Rights, and Neoliberalism." *Geoforum* 42, no. 1 (2011): 40–50. https://doi.org/10.1016/j.geoforum.2010.09.001.

Bullard, Robert D. *Confronting Environmental Racism: Voices from the Grassroots*. Boston: South End Press, 1983.

Burgess, H. K., L. B. DeBey, H. E. Froehlich, N. Schmidt, E. J. Theobald, A. K. Ettinger, and J. K. Parrish. "The Science of Citizen Science: Exploring Barriers to Use as a Primary Research Tool." *Biological Conservation* 208 (2017): 113–120.

Burke, Brian J., and Nik Heynen. "Transforming Participatory Science into Socioecological Praxis: Valuing Marginalized Environmental Knowledges in the Face of Neoliberalization of Nature and Science." *Environment and Society: Advances in Research* 5 (2014): 7–27. https://doi.org/10.3167/ares.2014.050102.

Chandler, Mark, Stan Rullman, Jenny Cousins, Nafeesa Esmail, Elise Begin, Gitte Venicx, Cristina Eisenberg, and Marie Studer. "Contributions to Publications and Management Plans from 7 Years of Citizen Science: Use of a Novel Evaluation Tool on Earthwatch-Supported Projects." *Biological Conservation* 208 (2017): 163–173.

Colborn, Theo, Carol Kwiatkowski, Kim Schultz, and Mary Bachran. "Natural Gas Operations from a Public Health Perspective." *Human and Ecological Risk Assessment* 17, no. 5 (2011): 1039–1056.

Corburn, Jason. *Street Science: Community Knowledge and Environmental Health Justice.* Cambridge, Mass.: MIT Press, 2005.

Craglia, Max, and Lea Shanley. "Data Democracy—Increased Supply of Geospatial Information and Expanded Participatory Processes in the Production of Data." *International Journal of Digital Earth* 8, no. 9 (September 2015): 679–693. https://doi.org/10.1080/17538947.2015.1008214.

Curtis, Vickie. "Motivation to Participate in an Online Citizen Science Game: A Study of Foldit." *Science Communication* 37, no. 6 (2015): 723–746. https://doi.org/10.1177/1075547015609322.

Daniels, George H. "The Pure-Science Ideal and Democratic Culture." *Science, New Series* 156, no. 3783 (June 30, 1967): 1699–1705.

Delborne, Jason A. "Transgenes and Transgressions: Scientific Dissent as Heterogeneous Practice." *Social Studies of Science* 38, no. 4 (2008): 509–541. https://doi.org/Doi10.1177/0306312708089716.

Desmarais, Annette-Aurélie. "The Vía Campesina: Consolidating an International Peasant and Farm Movement." *Journal of Peasant Studies* 29, no. 2 (2002): 91–124.

Deutsch, William, Laura Lhotka, and Sergio Ruiz-Cordova. "Group Dynamics and Resource Availability of a Long-Term Volunteer

Water-Monitoring Program." *Society & Natural Resources* 22, no. 7 (2009): 637–649. http://www.tandfonline.com/doi/abs/10.1080/08941920802078216.

Devictor, Vincent, Robert J. Whittaker, and Coralie Beltrame. "Beyond Scarcity: Citizen Science Programmes as Useful Tools for Conservation Biogeography." *Diversity and Distributions* 16, no. 3 (2010): 354–362.

DiChristopher, Tom. "US Shale Oil Will Surge to Nearly 7 Million Barrels a Day in April: Dept. of Energy Forecast." *CNBC*, March 13, 2018. https://www.cnbc.com/2018/03/13/us-shale-oil-will-surge-to-nearly-7-million-barrels-a-day-in-april.html.

Dickinson, Janis L., and Rick Bonney. *Citizen Science: Public Participation in Environmental Research.* Ithaca, N.Y.: Cornell University Press, 2012.

Dickinson, Janis L., Benjamin Zuckerberg, and David N. Bonter. "Citizen Science as an Ecological Research Tool: Challenges and Benefits." *Annual Review of Ecology, Evolution, and Systematics* 41, no. 1 (December 2010): 149–172. https://doi.org/10.1146/annurev-ecolsys-102209-144636.

Dillon, Justin, Robert B. Stevenson, and Arjen E. J. Wals. "Introduction to the Special Section Moving from Citizen to Civic Science to Address Wicked Conservation Problems. Corrected by Erratum 12844." *Conservation Biology* 30, no. 3 (June 2016): 450–455. https://doi.org/10.1111/cobi.12689.

Drori, Gili S., John W. Meyer, Francisco O. Ramirez, and Evan Schofer. *Science in the Modern World Polity: Institutionalization and Globalization.* Stanford: Stanford University Press, 2003.

Dufalla, Ken. "Bromides: Where Forth They Come." *Greene County Messenger*, March 2, 2012. http://www.heraldstandard.com/gcm/columns/natures_corner/bromides-where-forth-they-come/article_29d6632a-60e0-5c47-919c-3a0a05e2d105.

Eaton, Emily, and Abby Kinchy. "Quiet Voices in the Fracking Debate: Ambivalence, Nonmobilization, and Individual Action in Two Extractive Communities (Saskatchewan and Pennsylvania)." *Energy Research and Social Science* 20 (2016): 22–30. https://doi.org/10.1016/j.erss.2016.05.005.

Edwards, Marc A., and Siddhartha Roy. "Academic Research in the 21st Century: Maintaining Scientific Integrity in a Climate of Perverse Incentives and Hypercompetition." *Environmental Engineering Science* 34, no. 1 (2017): 51–61. https://doi.org/10.1089/ees.2016.0223.

Eisenberg, Ann M. "Beyond Science and Hysteria: Reality and Perceptions of Environmental Justice Concerns Surrounding Marcellus and Utica Shale Gas Development." *University of Pittsburgh Law Review* 77 (2015). https://doi.org/10.5195/lawreview.2015.396.

Eliasoph, Nina. *Making Volunteers: Civic Life after Welfare's End.* Princeton: Princeton University Press, 2011.

Eliasoph, Nina, and Paul Lichterman. "Making Things Political." In *Handbook of Cultural Sociology*, edited by John R. Hall, Laura Grindstaff, and Ming-Cheng Lo, 483–493. London: Routledge, 2010.

Elwood, Sarah. "Negotiating Knowledge Production: The Everyday Inclusions, Exclusions, and Contradictions of Participatory GIS Research." *Professional Geographer* 58, no. 2 (2006): 197–208. https://doi.org/10.1111/j.1467-9272.2006.00526.x.

Entrekin, Sally, Michelle Evans-White, Brent Johnson, and Elisabeth Hagenbuch. "Rapid Expansion of Natural Gas Development Poses a Threat to Surface Waters." *Frontiers in Ecology and the Environment* 9, no. 9 (November 2011): 503–511. https://doi.org/10.1890/110053.

Esteva, Gustavo, and Catherine Marielle. *Sin maíz no hay país.* Mexico City: Museo Nacional de Culturas Populares, 2003.

ETC Group. "Nine Mexican States Found to Be GM Contaminated." News release, 2003. http://www.etcgroup.org/en/materials/publications.html?id=145.

Ezcurra, Exequiel, and Jorge Soberón Mainero. "Evidence of Gene Flow from Transgenic Maize to Local Varieties in Mexico." In *LMOs and the Environment: Proceedings of an International Conference*, edited by OECD, 289–295. Paris: OECD, 2002.

Fernandez-Gimenez, Maria E., Heidi L. Ballard, and Victoria E. Sturtevant. "Adaptive Management and Social Learning in Collaborative and Community-Based Monitoring: A Study of Five Community-Based Forestry Organizations in the Western USA." *Ecology and Society* 13, no. 2 (2008): 4. http://www.mtnforum.org/sites/default/files/pub/4143.pdf.

Fitting, Elizabeth. "Importing Corn, Exporting Labor: The Neoliberal Corn Regime, GMOs, and the Erosion of Mexican Biodiversity." *Agriculture and Human Values* 23, no. 1 (2006): 15–26.

Frickel, Scott, Richard Campanella, and M. Bess Vincent. "Mapping Knowledge Investments in the Aftermath of Hurricane Katrina: A New Approach for Assessing Regulatory Agency Responses to Environmental Disaster." *Environmental Science & Policy* 12, no. 2 (2009): 119–133.

Frickel, Scott, Sahra Gibbon, Jeff Howard, Joanna Kempner, Gwen Ottinger, and David J. Hess. "Undone Science: Charting Social Movement and Civil Society Challenges to Research Agenda Setting." *Science, Technology & Human Values* 35, no. 4 (October 2009): 444–473. https://doi.org/10.1177/0162243909345836.

Frickel, Scott, and Kelly Moore, eds. *The New Political Sociology of Science: Institutions, Networks, and Power.* Madison: University of Wisconsin Press, 2005.

Funabashi, Harutoshi, Nobuko Iijima, and Koichi Hasegawa. *Kakunenryō-saikurushisetsuno shakaigaku* [Sociology of nuclear fuel recycle facilities]. Tokyo: Yuhikaku, 2012.

Gabrys, Jennifer, and Helen Pritchard. "Just Good Enough Data and Environmental Sensing: Moving beyond Regulatory Benchmarks toward Citizen Action." *International Journal of Spatial Data Infrastructures Research* 13 (2016): 4–14. https://doi.org/10.2902/1725-0463.2018.13.art2.

Gabrys, Jennifer, Helen Pritchard, and Benjamin Barratt. "Just Good Enough Data: Figuring Data Citizenships through Air Pollution Sensing and Data Stories." *Big Data & Society* 3, no. 2 (2016). https://doi.org/10.1177/2053951716679677.

Gittleman, Mara, Kelli Jordan, and Eric Brelsford. "Using Citizen Science to Quantify Community Garden Crop Yields." *Cities and the Environment (CATE)* 5, no. 1 (2012): 1–14.

Gomez, Rafael, and Morley Gunderson. "Volunteer Activity and the Demands of Work and Family." *Relations Industrielles / Industrial Relations* 58, no. 4 (2003): 573–589.

González, Roberto J. *Zapotec Science: Farming and Food in the Northern Sierra of Oaxaca.* Austin: University of Texas Press, 2001.

Gottlieb, Robert, and Anupama Joshi. *Food Justice.* Cambridge, Mass.: MIT Press, 2010.

Gouveia, Cristina, and Alexandra Fonseca. "New Approaches to Environmental Monitoring: The Use of ICT to Explore Volunteered Geographic Information." *GeoJournal* 72, no. 3–4 (2008): 185–197. https://doi.org/10.1007/s10708-008-9183-3.

Gustetic, Jenn, Kristen Honey, and Lea Shanley. "Open Science and Innovation: Of the People, by the People, for the People." Obama White House, 2015. https://obamawhitehouse.archives.gov/blog/2015/09/09/open-science-and-innovation-people-people-people.

Haklay, Muki. *Citizen Science and Policy: A European Perspective*. Vol. 4. Washington, D.C.: Woodrow Wilson Center, 2015.

———. "Citizen Science and Volunteered Geographic Information: Overview and Typology of Participation." In *Crowdsourcing Geographic Knowledge: Volunteered Geographic Information (VGI) in Theory and Practice*, edited by Daniel Sui, Sarah Elwood, and Michael Goodchild, 105–122. Dordrecht: Springer Netherlands, 2013.

Harding, Sandra. *Objectivity and Diversity: Another Logic of Scientific Research*. Chicago: University of Chicago Press, 2015.

———. *Science and Social Inequality: Feminist and Postcolonial Issues*. Champaign: University of Illinois Press, 2006.

———. *Whose Science? Whose Knowledge?* Ithaca, N.Y.: Cornell University Press, 1991.

Haro von Mogel, Karl. "Final Steps on the GMO Corn Experiment." Biology Fortified, data from August 20, 2017. https://www.biofortified.org/2017/08/final-steps-gmo-experiment/.

Haro von Mogel, Karl, and Anastasia Bodnar. "The GMO Corn Experiment." Experiment.com, data from March 30, 2018. https://experiment.com/projects/the-gmo-corn-experiment.

Harrison, Jill. "Parsing 'Participation' in Action Research: Navigating the Challenges of Lay Involvement in Technically Complex Participatory Science Projects." *Society & Natural Resources* 24, no. 7 (2011): 702–716. https://doi.org/10.1080/08941920903403115.

Hecht, Gabrielle. *Being Nuclear: Africans and the Global Uranium Trade*. Cambridge, Mass.: MIT Press, 2012.

Hemmi, Akiko, and Ian Graham. "Hacker Science versus Closed Science: Building Environmental Monitoring Infrastructure." *Information, Communication & Society* 17, no. 7 (2014): 830. https://doi.org/10.1080/1369118X.2013.848918.

Henriques, Gisele, and Raj Patel. *Policy Brief No. 7: Agricultural Trade Liberalization and Mexico*. Oakland, Calif.: Food First / Institute for Food and Development Policy, 2003.

Hess, David J. *Alternative Pathways in Science and Technology: Activism, Innovation, and the Environment in an Era of Globalization.* Cambridge, Mass.: MIT Press, 2007.

———. "The Potentials and Limitations of Civil Society Research: Getting Undone Science Done." *Sociological Inquiry* 79, no. 3 (August 2009): 306–327. https://doi.org/10.1111/j.1475-682X.2009.00292.x.

Higuchi, Kenji. *Yaminikesareru genpatuhibakusya* [Invisible exposed people]. Tokyo: Hachigatsu Shobo, 2011.

Hirakawa, Hideyuki, and Masashi Shirabe. "Rhetorical Marginalization of Science and Democracy: Politics in Risk Discourse on Radioactive Risks in Japan." In *Lessons from Fukushima: Japanese Case Studies on Science, Technology and Society*, edited by Yuko Fujigaki, 57–86. London: Springer International, 2015. https://doi.org/10.1007/978-3-319-15353-7_4.

Howard, Philip H. "Intellectual Property and Consolidation in the Seed Industry." *Crop Science* 55, no. 6 (2015): 2489–2495.

Ichinomiya, Yoshinari, Hiroaki Koide, Tomohiko Suzuki, and Takashi Hirose. *Genpatsusaikado no fukaiyami* [Problems of restarting the nuclear reactors]. Tokyo: Takarajimasha, 2012.

Iida, Tetsunari, Eisaku Sato, and Taro Kono. *Genshiryokumura wo koete* [Beyond the nuclear village]. Tokyo: NHK Books, 2011.

Irwin, Alan. *Citizen Science: A Study of People, Expertise and Sustainable Development.* London: Routledge, 1995.

Isin, E. F., and P. Nyers. *Routledge Handbook of Global Citizenship Studies.* London: Routledge, 2014.

Jackson, Robert B., Avner Vengosh, Thomas H. Darrah, Nathaniel R. Warner, Adrian Down, Robert J. Poreda, Stephen G. Osborn, Kaiguang Zhao, and Jonathan D. Karr. "Increased Stray Gas Abundance in a Subset of Drinking Water Wells near Marcellus Shale Gas Extraction." *Proceedings of the National Academy of Sciences of the United States of America* 110, no. 28 (July 9, 2013): 11250–11255. https://doi.org/10.1073/pnas.1221635110.

Jalbert, Kirk, Abby Kinchy, and Simona L. Perry. "Civil Society Research and Marcellus Shale Natural Gas Development: Results of a Survey of Volunteer Water Monitoring Organizations." *Journal of Environmental Studies and Sciences* 4, no. 1 (2014): 78–86. http://dx.doi.org/10.1007/s13412-013-0155-7.

Kaminer, Wendy. *Women Volunteering: The Pleasure, Pain, and Politics of Unpaid Work from 1830 to the Present*. New York: Anchor, 1984.

Katsuki, Keisuke. "Hosyanosokutei purojekuto Safecast" [Radiation measurement project, Safecast]. *Politas*, March 2015. http://politas.jp/features/4/article/354.

Kawamura, Minato. *Fukushimagenpatsu jinsaiki* [Fukushima nuclear reactor human disaster]. Tokyo: Gendaishokan, 2011.

Khoury, Lara, and Stuart Smyth. "Reasonable Foreseeability and Liability in Relation to Genetically Modified Organisms." *Bulletin of Science, Technology & Society* 27, no. 3 (2007): 215–232. https://doi.org/10.1177/0270467607300635.

Kimura, Aya H. "Citizen Monitoring in Japan: Spiderwort and Cherryblossoms." Article in preparation.

———. "Fukushima ETHOS: Post-disaster Risk Communication, Affect, and Shifting Risks." *Science as Culture* 27, no. 1 (2018): 98–117. https://doi.org/10.1080/09505431.2017.1325458.

———. *Hidden Hunger: Gender and the Politics of Smarter Foods*. Ithaca, N.Y.: Cornell University Press, 2013.

———. *Radiation Brain Moms and Citizen Scientists: The Gender Politics of Food Contamination after Fukushima*. Durham, N.C.: Duke University Press, 2016.

———. "Risk Communication under Post-feminism: Analysis of Risk Communication Programmes after the Fukushima Nuclear Accident." *Science Technology & Society* 21, no. 1 (2016): 24–41. https://doi.org/10.1177/0971721815622738.

Kinchy, Abby. *Seeds, Science, and Struggle: The Global Politics of Transgenic Crops*. Cambridge, Mass.: MIT Press, 2012.

Kinchy, Abby, Sarah Parks, and Kirk Jalbert. "Fractured Knowledge: Mapping the Gaps in Public and Private Water Monitoring Efforts in Areas Affected by Shale Gas Development." *Environment and Planning C: Government and Policy* 34, no. 5 (2016): 879–899. https://doi.org/10.1177/0263774X15614684.

Kinchy, Abby, and Simona L. Perry. "Can Volunteers Pick Up the Slack? Efforts to Remedy Knowledge Gaps about the Watershed Impacts of Marcellus Shale Gas Development." *Duke Environmental Law and Policy Journal* 22, no. 2 (2012): 303–339.

Kinchy, Abby, Kirk Jalbert, and Jessica Lyons. "What Is Volunteer Water
Monitoring Good For? Fracking and the Plural Logics of Participatory
Science." *Political Power and Social Theory* 27 (2014): 259–289. https://doi
.org/10.1108/S0198-871920140000027017.

Kleinman, Daniel Lee. *Impure Cultures: University Biology and the World of
Commerce*. Madison: University of Wisconsin Press, 2003.

Kleinman, Daniel Lee, and Abby Kinchy. "Against the Neoliberal Steamroller? The Biosafety Protocol and the Social Regulation of Agricultural
Biotechnologies." *Agriculture and Human Values* 24, no. 2 (2007): 195–206.

———. "Why Ban Bovine Growth Hormone? Science, Social Welfare, and
the Divergent Biotech Policy Landscapes in Europe and the United
States." *Science as Culture* 12, no. 3 (2003): 375–414.

Kloppenburg, Jack. *First the Seed: The Political Economy of Plant Biotechnology*.
Madison: University of Wisconsin Press, 2005.

Kobayashi, Tadashi. *Toransu saiensu no jidai* [The age of trans-science].
Tokyo: NTT Shuppan, 2007.

Koide, Hiroaki. *Genpatsu no uso* [Lies of the nuclear power plants]. Tokyo:
Fusosha, 2011.

———. *Hōshanoosen no genjitsu wo koete* [Beyond the reality of radiation
contamination]. Tokyo: Hokuto Shuppan, 1992.

Kuchinskaya, Olga. *The Politics of Invisibility: Public Knowledge about Radiation Health Effects after Chernobyl*. Cambridge, Mass.: MIT Press, 2014.

Kullenberg, Christopher, and Dick Kasperowski. "What Is Citizen Science?
A Scientometric Meta-analysis." *PLOS ONE* 11, no. 1 (2016): e0147152.
https://doi.org/10.1371/journal.pone.0147152.

Lave, Rebecca. "Neoliberalism and the Production of Environmental
Knowledge." *Environment and Society: Advances in Research* 3, no. 1
(December 18, 2012): 19–38. https://doi.org/10.3167/ares.2012.030103.

Lawrence, Anna. "'No Personal Motive?' Volunteers, Biodiversity, and the
False Dichotomies of Participation." *Ethics, Place & Environment* 9,
no. 3 (2006): 279–298. http://www.tandfonline.com/doi/abs/10.1080/
13668790600893319.

Leach, Melissa, Ian Scoones, and Brian Wynne. *Science and Citizens: Globalization and the Challenge of Engagement*. London: Zed Books, 2005.

Legere, Laura. "Citizen Training for Spotting Drilling Problems Criticized
by Natural Gas Industry." *Scranton Times-Tribune*, December 1, 2009.

Lewandowski, Eva J., and Karen S. Oberhauser. "Butterfly Citizen Science Projects Support Conservation Activities among Their Volunteers." *Citizen Science: Theory and Practice* 1, no. 6 (2016): 1–8. https://doi.org/10.5334/cstp.10.

Liévanos, Raoul S., Jonathan K. London, and Julie Sze. "Uneven Transformations and Environmental Justice: Regulatory Science, Street Science, and Pesticide Regulation in California." In *Technoscience and Environmental Justice: Expert Cultures in a Grassroots Movement*, edited by Gwen Ottinger and Benjamin Cohen, 201–228. Cambridge, Mass.: MIT Press, 2011.

Macey, Gregg P., Ruth Breech, Mark Chernaik, Caroline Cox, Denny Larson, Deb Thomas, and David O. Carpenter. "Air Concentrations of Volatile Compounds near Oil and Gas Production: A Community-Based Exploratory Study." *Environmental Health* 13, no. 1 (2014): 82. https://doi.org/10.1186/1476-069X-13-82.

Malin, Stephanie A., and Kathryn Teigen DeMaster. "A Devil's Bargain: Rural Environmental Injustices and Hydraulic Fracturing on Pennsylvania's Farms." *Journal of Rural Studies* 47, no. 2015 (2016): 278–290. https://doi.org/10.1016/j.jrurstud.2015.12.015.

Maloney, Kelly O., Sharon Baruch-Mordo, Lauren A. Patterson, Jean Philippe Nicot, Sally A. Entrekin, Joseph E. Fargione, Joseph M. Kiesecker, et al. "Unconventional Oil and Gas Spills: Materials, Volumes, and Risks to Surface Waters in Four States of the U.S." *Science of the Total Environment* 581/582 (2017): 369–377. https://doi.org/10.1016/j.scitotenv.2016.12.142.

Matz, Jacob Robert, Sara Wylie, and Jill Kriesky. "Participatory Air Monitoring in the Midst of Uncertainty: Residents' Experiences with the Speck Sensor." *Engaging Science, Technology, and Society* 3 (2017): 464. https://doi.org/10.17351/ests2017.127.

McCarthy, James. "Scale, Sovereignty, and Strategy in Environmental Governance." *Antipode* 37, no. 4 (September 2005): 731–753. https://doi.org/10.1111/j.0066-4812.2005.00523.x.

McCormick, Sabrina. "Transforming Oil Activism: From Legal Constraints to Evidenciary Opportunity." *Sociology of Crime Law and Deviance* 17 (2012): 113–131.

McNutt, Marcia. "The New Patrons of Research." *Science* 344, no. 6179 (2014): 9.

Meentemeyer, Ross K., Monica A. Dorning, John B. Vogler, Douglas Schmidt, and Matteo Garbelotto. "Citizen Science Helps Predict Risk of Emerging Infectious Disease." *Frontiers in Ecology and the Environment* 13, no. 4 (2015): 189–194. https://doi.org/10.1890/140299.

Michaels, Craig, James L. Simpson, and William Wegner. *Fractured Communities: Case Studies of the Environmental Impacts of Industrial Gas Drilling*. New York: Riverkeeper, 2010. https://www.riverkeeper.org/wp-content/uploads/2010/09/Fractured-Communities-FINAL-September-2010.pdf.

Mills, C. Wright. *The Sociological Imagination*. 40th anniversary ed. New York: Oxford University Press, 2000.

Mirowski, Philip. "Against Citizen Science." *Aeon*, no. 20 (November 2017). https://aeon.co/essays/is-grassroots-citizen-science-a-front-for-big-business.

Mooney, Chris. *The Republican War on Science*. New York: Basic Books, 2007.

Moore, Kelly, Daniel Lee Kleinman, David Hess, and Scott Frickel. "Science and Neoliberal Globalization: A Political Sociological Approach." *Theory and Society* 40, no. 5 (2011): 505–532.

Morita, Atsuro, Anders Blok, and Shuhei Kimura. "Environmental Infrastructures of Emergency: The Formation of a Civic Radiation Monitoring Map during the Fukushima Disaster." In *Nuclear Disaster at Fukushima Daiichi: Social, Political and Environmental Issues*, edited by Richard Hindmarsh, 78–96. New York: Routledge, 2013.

Muehlebach, Andrea. *The Moral Neoliberal: Welfare and Citizenship in Italy*. Chicago: University of Chicago Press, 2012.

Musick, Marc A., and John Wilson. *Volunteers: A Social Profile*. Bloomington: Indiana University Press, 2007.

Nash, Linda. *Inescapable Ecologies: A History of Environment, Disease, and Knowledge*. Berkeley: University of California Press, 2006.

National Research Council. *Learning Science in Informal Environments: People, Places, and Pursuits*. Washington, D.C.: National Academies Press, 2009.

Nerbonne, Julia Frost, and Kristen C. Nelson. "Volunteer Macroinvertebrate Monitoring in the United States: Resource Mobilization and

Comparative State Structures." *Society & Natural Resources* 17, no. 3 (September 2004): 817–839. https://doi.org/10.1080/08941920490493837.

Nerbonne, Julia Frost, Brad Ward, Ann Ollila, and Bruce Vondracek. "Effect of Sampling Protocol and Volunteer Bias When Sampling for Macro-invertebrates." *Journal of the North American Benthological Society* 27, no. 3 (2008): 640–646.

Newman, Greg, Don Zimmerman, Alycia Crall, Melinda Laituri, Jim Graham, and Linda Stapel. "User-Friendly Web Mapping: Lessons from a Citizen Science Website." *International Journal of Geographical Information Science* 24, no. 12 (November 26, 2010): 1851–1869. https://doi.org/10.1080/13658816.2010.490532.

Niedbala, Bob. "DEP Final Report on Ten Mile Creek Indicates No Dangers of Radioactivity." *Observer-Reporter* (Washington, Pa.), December 2, 2016. http://www.observer-reporter.com/20161202/dep_final_report_on_ten_mile_creek_indicates_no_dangers_of_radioactivity.

O'Rourke, Dara, and Gregg P. Macey. "Community Environmental Policing: Assessing New Strategies of Public Participation in Environmental Regulation." *Journal of Policy Analysis and Management* 22, no. 3 (2003): 383–414. https://doi.org/10.1002/pam.10138.

Osborn, Stephen G., Avner Vengosh, Nathaniel R. Warner, and Robert B. Jackson. "Methane Contamination of Drinking Water Accompanying Gas-Well Drilling and Hydraulic Fracturing." *Proceedings of the National Academy of Sciences of the United States of America* 108, no. 20 (May 17, 2011): 8172–8176. https://doi.org/10.1073/pnas.1100682108.

Ottinger, Gwen. "Buckets of Resistance: Standards and the Effectiveness of Citizen Science." *Science, Technology & Human Values* 35, no. 2 (June 12, 2010): 244–270. https://doi.org/10.1177/0162243909337121.

———. "Is It Good Science? Activism, Values, and Communicating Politically Relevant Science." *JCOM—Journal of Science Communication* 14, no. 2 (2015): C02.

———. *Refining Expertise: How Responsible Engineers Subvert Environmental Justice Challenges.* New York: New York University Press, 2013.

———. "Social Movement–Based Citizen Science." In *The Rightful Place of Science: Citizen Science*, edited by Darlene Cavalier and Eric B. Kennedy, 89–103. Tempe, Ariz.: Consortium for Science, Policy, and Outcomes, 2016.

Overdevest, Christine, Cailin Huyck Orr, and Kristine Stepenuck. "Volunteer Stream Monitoring and Local Participation in Natural Resource Issues." *Research in Human Ecology* 11, no. 2 (2004): 177–185.

Owen-Smith, Jason. "Commercial Imbroglios: Proprietary Science and the Contemporary University." In *The New Political Sociology of Science: Institutions, Networks, and Power,* edited by Scott Frickel and Kelly Moore, 63–90. Madison: University of Wisconsin Press, 2006.

Pearson, Thomas W. *When the Hills Are Gone: Frac Sand Mining and the Struggle for Community.* Minneapolis: University of Minnesota Press, 2017.

Peck, Jamie, and Adam Tickell. "Neoliberalizing Space." *Antipode* 34, no. 3 (June 2002): 380–404. http://www.blackwell-synergy.com/links/doi/10.1111/1467-8330.00247.

Pellow, David N. "Popular Epidemiology and Environmental Movements: Mapping Active Narratives for Empowerment." *Humanity & Society* 21, no. 3 (1997): 307–321. http://has.sagepub.com/content/21/3/307.short.

Penningroth, Stephen M., Matthew M. Yarrow, Abner X. Figueroa, Rebecca J. Bowen, and Soraya Delgado. "Community-Based Risk Assessment of Water Contamination from High-Volume Horizontal Hydraulic Fracturing." *New Solutions: A Journal of Environmental and Occupational Health Policy: NS* 23, no. 1 (January 1, 2013): 137–166. https://doi.org/10.2190/NS.23.1.i.

Perrow, Charles. *Normal Accidents: Living with High-Risk Technologies.* New York: Basic Books, 1984.

Perry, Simona L. "Development, Land Use, and Collective Trauma: The Marcellus Shale Gas Boom in Rural Pennsylvania." *Culture, Agriculture, Food and Environment* 34, no. 1 (June 26, 2012): 81–92. https://doi.org/10.1111/j.2153-9561.2012.01066.x.

Pfeffer, Max, and Linda Plummer Wagenet. "Volunteer Environmental Monitoring, Knowledge Creation and Citizen–Scientist Interaction." In *Sage Handbook on Environment and Society,* edited by Jules Pretty, Andrew S. Ball, Ted Benton, Julia S. Guivant, David R. Lee, David Orr, Max J. Pfeffer, and Hugh Ward, 235–249. London: Sage, 2007.

Pielke, Roger A. *The Honest Broker: Making Sense of Science in Policy and Politics.* Cambridge: Cambridge University Press, 2007.

Plantin, Jean-Christophe. "The Politics of Mapping Platforms: Participatory Radiation Mapping after the Fukushima Daiichi Disaster." *Media, Culture & Society* 37, no. 6 (2015): 904–921.

Pleasants, John M., and Karen S. Oberhauser. "Milkweed Loss in Agricultural Fields Because of Herbicide Use: Effect on the Monarch Butterfly Population." *Insect Conservation and Diversity* 6, no. 2 (March 2013): 135–144. https://doi.org/10.1111/j.1752-4598.2012.00196.x.

Pleasants, John M., Myron P. Zalucki, Karen S. Oberhauser, Lincoln P. Brower, Orley R. Taylor, and Wayne E. Thogmartin. "Interpreting Surveys to Estimate the Size of the Monarch Butterfly Population: Pitfalls and Prospects." Edited by Travis Longcore. *PLOS ONE* 12, no. 7 (July 14, 2017): e0181245. https://doi.org/10.1371/journal.pone.0181245.

Porticella, Norman, Susan Bonfield, Tony DeFalco, Ann Fumarolo, Cecilia Garibay, Eric Jolly, Laura Huerta Migus, et. al. *Promising Practices for Community Partnerships: A Call to Support More Inclusive Approaches to Public Participation in Scientific Research.* Washington, D.C.: Association of Science-Technology Centers, 2013.

Prainsack, Barbara, and Hauke Riesch. "Interdisciplinarity Reloaded: Drawing Lessons from 'Citizen Science.'" In *Investigating Interdisciplinary Collaboration: Theory and Practice*, edited by Scott Frickel, Albert Mathieu, and Barbara Prainsack, 151–164. New Brunswick, N.J.: Rutgers University Press, 2016.

Proctor, Robert N. *Value Free Science? Purity and Power in Modern Knowledge.* Boston: Harvard University Press, 1991.

Quarantelli, E. L., ed. *What Is a Disaster? Perspectives on the Question.* New York: Taylor & Francis, 1998.

Quist, David, and Ignacio H. Chapela. "Transgenic DNA Introgressed into Traditional Maize Landraces in Oaxaca, Mexico." *Nature* 414, no. 6863 (2001): 541–543.

Radder, Hans. *The Commodification of Academic Research: Science and the Modern University.* Pittsburgh: Pittsburgh University Press, 2010.

Rahman, Serina. *Johor's Forest City Faces Critical Challenges.* Singapore: ISEAS-Yusof Ishak Institute, 2017.

Ratto, Matt. "Critical Making: Conceptual and Material Studies in Technology and Social Life." *Information Society* 27, no. 4 (2011): 252–260.

Rip, Arie. "The Past and Future of RRI." *Life Sciences, Society and Policy* 10, no. 1 (December 2014): 17. https://doi.org/10.1186/s40504-014-0017-4.

Rubight, Sam. "Help Us to Track Oil Trains in Pittsburgh and Beyond." FracTracker Alliance, October 15, 2014. https://www.fractracker.org/2014/10/track-oil-trains/.

Sattler, Franziska, and Claudia Göbel. "Summary of Results from Our First #CitSciChat on CS and Responsible Research and Innovation." European Citizen Science Association (ECSA). N.d. https://ecsa.citizen-science.net/blog/summary-results-our-first-citscichat-cs-and-responsible-research-and-innovation.

Schurman, Rachel A., and Dennis Doyle Takahashi Kelso. *Engineering Trouble: Biotechnology and Its Discontents.* 1st ed. Berkeley: University of California Press, 2003.

Schwartz, Mark D., Julio L. Betancourt, and Jake F. Weltzin. "From Caprio's Lilacs to the USA National Phenology Network." *Frontiers in Ecology and the Environment* 10, no. 6 (2012): 324–327.

Secretariat of the Commission for Environmental Cooperation. *Maize & Biodiversity: The Effects of Transgenic Maize in Mexico.* Montreal: Communications Department of the Commission for Environmental Cooperation, 2004.

Serrano Sanz, Fermín, Teresa Holocher-Ertl, Barbara Kieslinger, Francisco Sanz García, and Cândida G. Silva. *White Paper on Citizen Science for Europe.* Brussels: Commission européenne, 2014.

Shapiro, Nicholas, Nasser Zakaria, and Jody A. Roberts. "A Wary Alliance: From Enumerating the Environment to Inviting Apprehension." *Engaging Science, Technology, and Society* 3 (2017): 575–602.

Shirk, Jennifer, Heidi L. Ballard, Candie C. Wilderman, Tina Phillips, Andrea Wiggins, Rebecca Jordan, Ellen McCallie, et al. "Public Participation in Scientific Research: A Framework for Deliberate Design." *Ecology and Society* 17, no. 2 (2012): 29. https://doi.org/10.5751/ES-04705-170229.

Shiva, Vandana. *Monocultures of the Mind: Perspectives on Biodiversity and Biotechnology.* London: Zed Books, 1993.

Shrader-Frechette, Kristin. *What Will Work: Fighting Climate Change with Renewable Energy, Not Nuclear Power.* Oxford: Oxford University Press, 2011.

Slater, David H., Rika Morioka, and Haruka Danzuka. "Micro-politics of Radiation." *Critical Asian Studies* 46, no. 3 (July 2014): 485–508.

Smith, John N., Vincent Rossi, Ken O. Buesseler, Jay T. Cullen, Jack Cornett, Richard Nelson, Alison M. Macdonald, Marie Robert, and Jonathan Kellogg. "Recent Transport History of Fukushima Radioactivity in the Northeast Pacific Ocean." *Environmental Science and Technology* 51, no. 18 (2017): 10494–10502. https://doi.org/10.1021/acs.est.7b02712.

Soeder, Daniel J., and William M. Kappel. *Water Resources and Natural Gas Production from the Marcellus Shale.* Fact Sheet 2009-3032. Washington, D.C.: U.S. Department of Interior, 2009.

Soleri, Daniela, Jonathan W. Long, Mónica D. Ramirez-Andreotta, and Rose Eitemiller. "Finding Pathways to More Equitable and Meaningful Public-Scientist Partnerships." *Citizen Science: Theory and Practice* 1, no. 1 (2016): 1–11. http://doi.org/10.5334/cstp.46.

Stacy, Andrew. "Additional WVU Testing Confirms Acceptable Levels of Radioactivity in Drinking Water at Clyde Mine." *Three Rivers Quest News*, August 27, 2015. http://3riversquest.org/additional-wvu-testing -confirms-radioactivity-below-drinking-water-standards-at-clyde -mine/.

Stedman, Richard, Brian Lee, Kathryn Brasier, Jason L. Weigle, and Francis Higdon. "Cleaning Up Water? Or Building Rural Community? Community Watershed Organizations in Pennsylvania." *Rural Sociology* 74, no. 2 (October 22, 2009): 178–200. https://doi.org/10.1111/j.1549-0831 .2009.tb00388.x.

Steinhoff, Patricia. "Memories of New Left Protest." *Journal of the German Institute for Japanese* 25, no. 2 (2013): 127–165.

Steneck, Nicholas. "Responsible Advocacy in Science: Standards, Benefits, and Risks." American Association for the Advancement of Science, 2011. https://www.aaas.org/resources/report-responsible-advocacy-science -standards-benefits-and-risks.

Stepenuck, Kristine F., and Linda T. Green. "Individual- and Community-Level Impacts of Volunteer Environmental Monitoring: A Synthesis of Peer-Reviewed Literature." *Ecology and Society* 20, no. 3 (2015): 19. https://doi.org/10.5751/ES-07329-200319.

Sugiura, Masakazu. "Heisei 26 nendo Kankyōshō oyobi genshiryokuki-seikankeiyosan no gaiyo" [Overview of budget for the Ministry of

Environment and Nuclear Power Regulations for FY 2014]. *Rippo to chosa* 349 (2014): 134–136.

Suryanarayanan, Sainath, and Daniel Lee Kleinman. *Vanishing Bees: Science, Politics, and Honeybee Health*. New Brunswick, N.J.: Rutgers University Press, 2016.

Suryanarayanan, Sainath, Daniel Lee Kleinman, Claudio Gratton, Amy Toth, Christelle Guedot, Russell Groves, John Piechowski, et al. "Collaboration Matters: Honey Bee Health as a Transdisciplinary Model for Understanding Real-World Complexity." *BioScience* 68, no. 12 (2018): 990–995.

Swanson, Alexandra, Margaret Kosmala, Chris Lintott, and Craig Packer. "A Generalized Approach for Producing, Quantifying, and Validating Citizen Science Data from Wildlife Images." *Conservation Biology* 30, no. 3 (June 2016): 520–531. https://doi.org/10.1111/cobi.12695.

Szasz, Andrew. "Progress through Mischief: The Social Movement Alternative to Secondary Associations." *Politics & Society* 20, no. 4 (1992): 521–528.

Taebi, Behnam, A. Correlje, E. Cuppen, M. Dignum, and U. Pesch. "Responsible Innovation as an Endorsement of Public Values: The Need for Interdisciplinary Research." *Journal of Responsible Innovation* 1, no. 1 (2014): 118–124.

Takagi, Jinzaburo. *Shimin kagakushatoshite ikiru* [Life as a citizen scientist]. Vol. 1. Tokyo: Nanatsumorishokan, 2002.

Taylor, Michael R., and Jody S. Tick. *The StarLink Case: Issues for the Future*. Washington, D.C.: Resources for the Future / Pew Initiative on Food and Biotechnology, 2001.

Theobald, E. J., A. K. Ettinger, H. K. Burgess, L. B. DeBey, N. R. Schmidt, H. E. Froehlich, C. Wagner, et al. "Global Change and Local Solutions: Tapping the Unrealized Potential of Citizen Science for Biodiversity Research." *Biological Conservation* 181 (2015). https://doi.org/10.1016/j.biocon.2014.10.021.

Thomas, Deborah. "Living with Oil and Gas and Practicing Community Conducted Science." *Engaging Science, Technology, and Society* 3 (2017): 613–618. https://doi.org/10.17351/ests2017.131.

Toerpe, Kathleen. "The Rise of Citizen Science." *Futurist* 47, no. 4 (2013): 25–30.

Trumbull, Deborah J., Rick Bonney, Derek Bascom, and Anna Cabral. "Thinking Scientifically during Participation in a Citizen-Science

Project." *Science Education* 84 (2000): 265–275. https://doi.org/10.1002/
(SICI)1098-237X(200003)84:2<265::AID-SCE7>3.3.CO;2-X.

Ureta, Sebastián. "Baselining Pollution: Producing 'Natural Soil' for an Environmental Risk Assessment Exercise in Chile." *Journal of Environmental
Policy & Planning* 7200 (November 2017): 1–14. https://doi.org/10.1080/
1523908X.2017.1410430.

U.S. Environmental Protection Agency. *Surface Water Monitoring: A Framework for Change.* Washington, D.C.: USEPA, 1987.

———. *Volunteer Water Monitoring: A Guide for State Managers.* Washington,
D.C.: USEPA, 1990.

Vetter, Jeremy. "Introduction: Lay Participation in the History of Scientific
Observation." *Science in Context* 24, no. 2 (April 28, 2011): 127–141. https://
doi.org/10.1017/S0269889711000032.

Vicens, Natasha. "DEP's Testing Methods for Radiation in a PA Creek
Questioned." *PublicSource,* July 30, 2015. http://publicsource.org/deps
-testing-methods-for-radiation-in-a-pa-creek-questioned/.

———. "How One Resident near Fracking Got the EPA to Pay Attention to Her Air Quality." *PublicSource,* December 15, 2016. https://www
.publicsource.org/how-one-resident-near-fracking-got-the-epa-to-pay
-attention-to-her-air-quality/.

———. "PA DEP Finds Safe Radioactivity Levels in Greene County
Creek." *PublicSource,* December 17, 2015. http://publicsource.org/pa-dep
-finds-safe-radioactivity-levels-in-greene-county-creek/.

Wiggins, Andrea, and Kevin Crowston. "From Conservation to Crowdsourcing: A Typology of Citizen Science." *Proceedings of the Annual
Hawaii International Conference on System Sciences* (2011): 1–10. https://
doi.org/10.1109/HICSS.2011.207.

Wilderman, Candie, and Jinnieth Monismith. "Monitoring Marcellus: A
Case Study of a Collaborative Volunteer Monitoring Project to Document the Impact of Unconventional Shale Gas Extraction on Small
Streams." *Citizen Science: Theory and Practice* 1, no. 1 (May 20, 2016).
https://doi.org/10.5334/cstp.20.

Williams, Robert W. "Environmental Injustice in America and Its Politics
of Scale." *Political Geography* 18, no. 1 (January 1999): 49–73. http://dx.doi
.org/10.1016/S0962-6298(98)00076-6.

Wilmsen, Carl, William Elmendorf, Larry Fisher, Jacquelyn Ross, Brinda Sarathy, and Gail Wells, eds. *Partnerships for Empowerment: Participatory Research for Community-Based Natural Resource Management*. London: Earthscan, 2008. https://doi.org/10.4324/9781849772143.

Wilson, Diane. "Community Based Water Monitoring and beyond, a Case Study: Pennsylvania." *Proceedings of the Water Environment Federation* (2002): 1025–1036. http://www.ingentaconnect.com/content/wef/wefproc/2002/00002002/00000002/art00063.

Wisman, Jon D. "Inequality, Social Respectability, Political Power, and Environmental Devastation." *Journal of Economic Issues* 45, no. 4 (2011): 877–900.

Wisner, Ben, Piers Blaikie, Terry Cannon, and Ian Davis. *At Risk: Natural Hazards, People's Vulnerability and Disasters*. London: Routledge, 2004.

Wofford, Pamela, Randy Segawa, and Jay Schreider. "Pesticide Air Monitoring in Parlier, CA." California Department of Pesticide Regulation, December 2009.

Wylie, Sara. *Fractivism: Corporate Bodies and Chemical Bonds*. Durham, N.C.: Duke University Press, 2018.

Wylie, Sara, Kirk Jalbert, Shannon Dosemagen, and Matt Ratto. "Institutions for Civic Technoscience: How Critical Making Is Transforming Environmental Research." *Information Society* 30, no. 2 (2014): 116–126. https://doi.org/10.1080/01972243.2014.875783.

Wylie, Sara, Elisabeth Wilder, Lourdes Vera, Deborah Thomas, and Megan McLaughlin. "Materializing Exposure: Developing an Indexical Method to Visualize Health Hazards Related to Fossil Fuel Extraction." *Engaging Science, Technology, and Society* 3 (2017): 426. https://doi.org/10.17351/ests2017.123.

Yagi, Tadashi. *Genpatsuwa sabetsude ugoku* [Nuclear energy is rooted in discrimination]. Tokyo: Akashi shoten, 2011.

Young, Iris Marion. *Inclusion and Democracy*. Cambridge: Oxford University Press, 2000.

———. "Polity and Group Difference: A Critique of the Ideal of Universal Citizenship." *Ethics* 99 (1989): 250–274.

Zilliox, Skylar, and Jessica M. Smith. "Colorado's Fracking Debates: Citizen Science, Conflict and Collaboration." *Science as Culture* 5431 (2018): 1–21. https://doi.org/10.1080/09505431.2018.1425384.

Index

Page numbers in *italics* refer to figures.

About the Authors

AYA H. KIMURA is an associate professor of sociology at the University of Hawai'i–Mānoa. She completed her MA in environmental studies at Yale University and her PhD in sociology at the University of Wisconsin-Madison. Her books include *Radiation Brain Moms and Citizen Scientists: The Gender Politics of Food Contamination after Fukushima* (recipient of the Rachel Carson Book Award from the Society for Social Studies of Science), *Hidden Hunger: Gender and the Politics of Smarter Foods* (recipient of the Outstanding Scholarly Achievement Award from the Rural Sociological Society), and *Food and Power: Visioning Food Democracy in Hawai'i* (coeditor).

ABBY KINCHY studies how social movements, such as the alternative agriculture movement and antifracking activists, resist and shape the development of controversial technologies. She is the author of *Seeds, Science, and Struggle: The Global Politics of Transgenic Crops* and coeditor of *Controversies in Science and Technology: From Maize to Menopause*. Kinchy completed her PhD in sociology at the University of Wisconsin–Madison and is currently an associate professor in the Science and Technology Studies Department at Rensselaer Polytechnic Institute. In addition to her research on science, technology, and social movements, Kinchy is a coorganizer of STS Underground, a research network that advances social science research on the technoscientific dimensions of mining, burial, and other forms of subterranean exploration.